SMART BORDERS, DIGITAL IDENTITY AND BIG DATA

How Surveillance Technologies Are Used Against Migrants

Emre Eren Korkmaz

First published in Great Britain in 2025 by

Bristol University Press
University of Bristol
1–9 Old Park Hill
Bristol
BS2 8BB
UK
t: +44 (0)117 374 6645
e: bup-info@bristol.ac.uk

Details of international sales and distribution partners are available at
bristoluniversitypress.co.uk

© Bristol University Press 2025

British Library Cataloguing in Publication Data
A catalogue record for this book is available from the British Library

ISBN 978-1-5292-3350-6 hardcover
ISBN 978-1-5292-5389-4 paperback
ISBN 978-1-5292-3351-3 ePub
ISBN 978-1-5292-3352-0 ePdf

The right of Emre Eren Korkmaz to be identified as the author of this work has been asserted by him in accordance with the Copyright, Designs and Patents Act 1988.

All rights reserved: no part of this publication may be reproduced, stored in a retrieval system, or transmitted in any form or by any means, electronic, mechanical, photocopying, recording, or otherwise without the prior permission of Bristol University Press.

Every reasonable effort has been made to obtain permission to reproduce copyrighted material. If, however, anyone knows of an oversight, please contact the publisher.

The statements and opinions contained within this publication are solely those of the author and not of the University of Bristol or Bristol University Press. The University of Bristol and Bristol University Press disclaim responsibility for any injury to persons or property resulting from any material published in this publication.

Bristol University Press works to counter discrimination on grounds of gender, race, disability, age and sexuality.

Cover design: Lyn Davies Design
Front cover image: Scharfsinn/ Alamy Stock Photo

To my dear wife Meltem and
kids Selen and Alp Efe

Contents

List of Abbreviations		vi
About the Author		vii
Acknowledgements		viii
Introduction: Canaries in the Coal Mine		1
1	Migration and (Surveillance) Capitalism	16
2	Migration and (Big) Data Analysis	46
3	Smart Borders	68
4	Digital Identity and Surveillance Capitalism	87
Conclusion: How Can We Resist?		116
Notes		133
References		135
Index		146

List of Abbreviations

AI	artificial intelligence
ANPR	automatic number plate recognition
Avatar	automated virtual agent for truth assessments in real time
CDR	call detail record
CIA	Central Intelligence Agency
DHS	United States Department of Homeland Security
ESA	European Space Agency
EU	European Union
EUAA	European Union Agency for Asylum
Eurodac	European dactyloscopy (EU fingerprint database)
Eu-LISA	EU Agency for the Operational Management of Large-Scale IT Systems in the Area of Freedom, Security and Justice
GDPR	general data protection regulation
GSMA	GSM Association (Groupe Spécial Mobile/Global System for Mobile Communications)
HART	Homeland Advanced Recognition Technology System
İBB	Istanbul Metropolitan Municipality
ID	identity
IFRC	International Federation of Red Cross and Red Crescent Societies
IOM	International Organization for Migration
NGO	non-governmental organization
SDG	(UN's) Sustainable Development Goals
SME	small- or medium-sized enterprise
UK	United Kingdom of Great Britain and Northern Ireland
UKRI	UK Research and Innovation
UN	United Nations
UNHCR	United Nations High Commissioner for Refugees
US	United States (of America)
WFP	World Food Programme

About the Author

Emre Eren Korkmaz worked at the Oxford Department of International Development (ODID) between 2016 and 2023 (for the first two years as a British Academy Fellow and then as Lecturer on the MSc in Migration Studies programme). He was also Junior Research Fellow at St Edmund Hall (2017–2020) and an affiliate of the Centre for Technology and Global Affairs (2018–2019). In October 2023, he joined University of Oxford's Centre on Migration, Policy and Society (COMPAS) as a research affiliate.

He gained a PhD from Istanbul University's International Relations PhD Programme in June 2016. He is driven by a passion to shed light on the social and political impact of technological innovation, particularly surveillance technologies in migration management, border security and humanitarianism.

Acknowledgements

I would like to wholeheartedly thank my dear wife Meltem for encouraging me to write this book and my lovely kids Selen and Alp Efe for allowing me to focus on it without creating so much trouble.

I would also like to thank Michael King, my student and supervisee at MSc in Migration Studies in Oxford, for his editorial support. His reviews and editorial suggestions had a crucial impact on the finalization of this book.

I am profoundly appreciative of the mentorship and support of Professor Ruben Andersson at Oxford. He guided me during my academic career from 2016 and supported and encouraged me in research, publications and teaching. I feel fortunate to work with many esteemed migration scholars – Professor Biao Xiang, Professor Robin Cohen, Professor Mathias Czaika, Professor Carlos Vargas-Silva and Professor Nicholas Van Hear, among others, at the Oxford Department of International Development and COMPAS. I also would like to thank Professor Diego Sánchez-Ancochea, head of ODID, and Professor Xiaolan Fu for their guidance and support.

I am delighted to receive very constructive reviews from my dear colleagues Dr Turkay Zengin, Ayşe Acar, Dr Nevzat Evrim Önal, Dr Samuel Singler, Verda Yüceer, Thomas von Davier, Merve Korkmaz, Dr Melih Yeşilbağ, Mia Haas-Goldberg, Aiden Slavin and my students on the 'New Technologies and People on the Move' course in Oxford.

Last but not least, I would like to thank Paul Stevens and Georgina Bolwell from Bristol University Press, who patiently guided me through rigorous review processes to create this book.

Introduction: Canaries in the Coal Mine

We are living in a time when technological transformation permeates all areas of life and affects radical change. Private companies, state actors and academic institutions are in fierce competition among themselves, fostering rapid progress in fields such as artificial intelligence algorithms, digital platforms, the internet of things, blockchain and quantum computers, and, as a result, new products are being produced on a daily basis. The future trajectory of this change, the problems it creates, the inequalities it generates, and the winners and losers of this process, among many other issues, have become significant topics of interest.

On the one hand, more and more people are enjoying the benefits of new technologies as they become cheaper and widely accessible. We have become better informed about the world, have more opportunities to socialize, embrace new business opportunities and generally navigate the world with confidence. We rely more than ever on our phones and computers to live, stress about our batteries draining and feel compelled to check our devices throughout the day. On the other hand, trust in traditional institutions has been shaken because of disinformation, fake news, algorithmic discrimination and cybersecurity concerns.

It is in the context of the issues described in this book that migration and border management are such key areas of investigation. Indeed, it is now much more difficult to come up with alternative legal solutions and identify or prevent discrimination or errors with potentially lethal consequences for migrants or asylum seekers. This is partly a matter of risks and problems brought about by automating a corrupt and discriminatory system.

While there are arrays of technological products that we cannot see (satellites), notice (internet of things, mobile phone social media analysis), influence (machine learning) or even understand (blockchain) which improve our lives, these coexist with technologies that make life more difficult; hampering and ignoring our ability to assert autonomy and make decisions. Indeed, there is a deep anxiety regarding our world in 20 years' time, as a result of technological advances and transformation – often reflected in utopian or, more commonly, dystopian works of art and media.

Governments and lawmakers are trying to enact laws and regulations regarding these issues and provide a legal framework for the changes. Labour unions are seeking to understand how they can react to new types of employment made possible by digital platforms, and intervene in changes in the workplace and algorithmic decision-making processes. While often utilizing technologies to reach wider audiences, social and political movements are concerned by the increasing surveillance and oppression associated with technological advancement. Small and medium enterprises are trying to navigate rapid monopolization in technology-generating sectors and the colossal power amassed by a small number of tech giants.

Technological solutionism – the optimistic view that societal problems can be solved by technology – is increasingly challenged by ethical concerns and direct observations and experience of the reproduction and deepening of structural inequalities. Developments in war technologies – which are widely tested on asylum-seekers and irregular migrants – for example regarding autonomous weapons and robot soldiers, drones and cyberattacks, have made wars and conflict more destructive, widespread and long term (Constantin, 2016; Kello, 2019); have shown that the destruction and horrors brought about by future conflicts would be much greater; and, increasingly, have given rise to demands to make greater efforts to protect peace and rein in producers of war technologies.

Therefore, this is not just about the competition among engineers, academics and experts – who have technical mastery over developments in the field of technology – to develop better, more effective and cheaper products. Rather, this is a field that, because it affects the daily lives and future visions of all segments of society, we all must think about, try to understand and influence. This, inherently, then, is a political conflict. The world we will inhabit in 20 years' time will be shaped by the outcome of this conflict. This book contends that migration management, border security and humanitarian emergencies are significant fields in which technology, military and financial corporations test and develop new surveillance technologies. Examining these speculative pilot projects and applications would provide sufficient knowledge, experience and understanding of utopian and dystopian images of the future. For instance, we have many cases from which to predict what would happen if an artificial general intelligence – one that is smarter and more capable than humans – were to emerge.

Current technological developments are sufficient to contribute to a more egalitarian social order (Fuchs, 2019); one that improves the living conditions of all people, lowers working hours, supports decision-makers to ensure that everyone receives fair compensation in return for their work and can travel all around the world to meet and interact with other people without linguistic or geographical barriers. Alternatively, they could also be used to establish a dictatorship that serves the interests of a small minority, oppresses and persecutes large social strata, and is based on blatant exploitation and extreme violence.

The current capitalist system is arguably headed in the dystopian, dictatorial direction. This shows that an approach to using technological products with only limited comprehension of the opportunities and contradictions they offer is not sufficient. It is vital to perceive this process as a field of political conflict and, accordingly, ask the question: 'What is the world that we want to live in?'

This is the reason why governments, legislatures, bureaucracies, various mass organizations, interest groups, trade bodies and private companies are trying, on the one hand, to understand the process, and on the other, to develop alternative methods and solutions rapidly and in fierce competition with one another.

More time to discussing these issues is given in the following chapters. To underline the political aspect of the issue one more time, political/societal problems can only have political solutions; technological products alone are not capable of providing solutions, and, more often than not, they would end up reinforcing existing inequalities, imbalances and injustices. This is why many researchers and NGOs studying border and migration management technologies call for outright bans on some technological solutions – because they bring more risks and deaths (Achiume, 2020; Privacy International, 2021).

This book does not limit its scope to applications of technology. At the end of the day, the technological developments being talked about are means of production of the corporations that commercialize them. An algorithm, robot or drone does not have an intention or consciousness of its own. These tools are limited by and accountable only to the capitalists that turn them into a means of production, establish private ownership over them, generate value out of them and decide which are to be produced. Therefore, when it comes to the production, development and commercialization of these tools, it would be better to focus on the interests and tendencies of those who establish private ownership over them. For example, in examining the services and products of companies such as Facebook/Meta, Google and Amazon, we cannot disregard the capitalists that own these firms and their relationships with shareholders, corporate managers and governments. These are organizations whose impacts and effects prevail far further than their stated products as, for instance, 'public spaces that allow people to socialize and improve democracy', 'platforms that provide search engine services and allow people to access information that they need', or 'solution partners that provide flexible employment to people and allow them to establish their own businesses'.

This book will focus on the corporations which produce surveillance tech commodities in migration and border management. However, it examines this process from an under-studied perspective in this field. Each commodity has a use-value and exchange-value.[1] While exchange-value is essential for corporations to commercialize these products, the use-value is significant for their clients, such as governments, UN agencies, humanitarian organizations and other corporations. Most academic studies and NGO reports criticize clients about the consequences of the use-value of surveillance technologies (Privacy International, 2016, 2020, 2021, 2022). However, this book will discuss the exchange-value to shed light on the reasons for technology, military and financial corporations to develop and test surveillance technologies with the support of governments of advanced capitalist states, who join this process as funders and clients.

Therefore, this book, which discusses technology in the context of migration and border management, must also make observations, draw conclusions and develop an understanding of the tendencies and contradictions of the owners of these means of production, as well as the tendencies of the capitalist system. Many concrete examples are given throughout the book, and you will see that the field of border management is not limited to migrants: it can have repercussions for the entire society and provides an important forecast to the overall direction, trajectory and contradictions of capitalism.

As Ruben Andersson (2022) rightly mentions, 'migrants may be the canaries in the coal mine of a much wider turn toward intimate exploitations of life itself for economic and political purposes'.

Reformist and structuralist approaches

This book divides critical approaches on technological transformation into two broad categories. One is reform-oriented to explore solutions to defined problems, whereas the other employs structural analysis to examine the direction of and changes in capitalism.

The reform-oriented approach evaluates the negative effects of certain actions taken by authorities or private companies on various social groups, and offers solutions. For example, there have been many studies on how the AI algorithms used by US police to identify suspects and predict and prevent crime, or algorithms that assist judges in their sentencing decisions, generate discrimination against minorities, most commonly Black people. As a result, many organizations are fighting for reform in this area (O'Neil, 2017; Noble, 2018; Milivojevic, 2021). In another example, an algorithm was used in the UK in 2021 to lower the marks given by teachers for high school exams – which were not held in traditional classroom environments because of the COVID-19 pandemic – in order to bring grades closer to the historical averages of individual schools. Students protested when it was evidenced that the algorithm lowered the grades of those attending state schools and residing in poor areas to a larger extent compared with more privileged students and private schools (Kolkman, 2020).

Many analyses have been made of the reasons why minorities and the working classes, who face structural oppression, exploitation and discrimination, are more negatively affected by the outcomes of these AI algorithms. In broad terms, it is because the AI algorithms in question are trained on existing data, and the working principles are designed by their private owners. Therefore, when they are asked to make a decision, their decisions are based on this training and development process. Consider, for example, sociological and political explanations of why Black people are over-represented among the prison population in the US due to the oppression and discrimination historically faced by Black people in that country. Without this critical knowledge, an algorithm tasked with identifying a criminal between two people – one who is Black and the other white – will incorporate historically tainted and unnuanced data sets to charge the former over the latter (O'Neil, 2017; Noble, 2018).

Similarly, it is often the case that private schools in the UK produce students with higher grades and send more of their graduates on to higher education. When you feed this information into an algorithm, because the algorithm would not know about the academic achievement levels and performances of individual students, it only makes sense for the algorithm to calculate the historical average of a given school and lower these students' marks to a larger extent. However, treating these outcomes as valid, neutral or scientific does not make sense.

The reform-oriented approach thus identifies current inequalities and injustices and makes efforts to offer solutions. Many academics, NGOs, trade bodies established by private companies and multi-stakeholder initiatives are making efforts to ensure that algorithms are trustworthy and explainable. Various methods are proposed, from adopting a participatory attitude in the design stage to ensuring the quality and diversity of the data sets used to train an algorithm (Floridi, 2007; Kelly, 2022).

In a well-known example of gender-based professional discrimination, if you type in a translation app '*O doktordur*', in Turkish, which does not have separate male and female pronouns, you get the translation '*He* is a doctor', while '*O hemşiredir*' is translated to '*She* is a nurse'. This is relatively easy to solve. In a distinctly different context – the case of autonomous weapons – AI, left unsupervised, can perceive non-existent risks because of a system error or miscalculation and start a war. One solution is to implement human supervision on final decisions to action the risk assessments generated by AI (Kello, 2019).

In many workplaces, if an employee faces discrimination because of an algorithm, they can defend their rights with the assistance of labour unions and through legal action. In the case of the gig economy on digital platforms such as Uber, Getir and Amazon, on the other hand, these 'self-employed' workers lack job security and are denied many benefits such as pensions and sick pay, and there are campaigns and legislative efforts under way to get digital platforms recognized as employers, given their demands on, control over and limitations for workers, and to require them to provide all the benefits that employees are due by law, such as those regarding minimum wage, social-security contributions and unionization. These campaigns, collective agreements, legislations and other safety measures aim to reform the system in order to overcome technological challenges.

Another approach to technological transformation focuses on structural aspects, studies the big picture and explores how technological developments affect capitalism, intensify exploitation and update labour relations and class positionings. The gig economy, for example, is studied in the context of precarious work. Autonomous weapons and war technologies are examined in the context of competition and division of markets between capitalist states or powers. Robots and autonomous tools are discussed in the context of unemployment and the future of the working class, and analyses of the monopolization of technology companies focus on the position of start-ups, other small enterprises and poor or small nations vis-à-vis these companies, utilizing concepts such as (neo)colonialism, dependency and imperialism (Fuchs, 2019).

Within the framework of the critical approach adopted in this book, I will make frequent use of reformist approaches, citing case studies and sector-specific studies, as well as a broad anti-capitalist analysis to address structural developments. I will also highlight the connections between surveillance technologies used in migration and border management with reference to examples from other literature, such as that on work, war technologies or future wars, surveillance and crime prediction and prevention. In addition to my own fieldwork research activities, I will utilize the increasing number of critical reports, books and articles on migration and border management published in recent years. I will analyse the causes and actual or potential consequences of technologies developed for and used – or, more commonly, piloted – in the field of migration, as well as their effects on other fields, and also discuss them in the context of the overall trajectory of capitalism. In doing so, I will build upon and contribute to the concept of surveillance capitalism.

Thus, the relevant audience of this book is not limited to people who work in the field of migration or are interested in issues of migrants. It also speaks to people who analyse the consequences of technology in different disciplines and those interested in the overall trajectory of capitalism. With concrete examples, this book shows that many issues discussed in other disciplines have become lived experiences in the field of migration and border management, attracting the attention of researchers from other fields.

It is hoped that this book will also be of interest in terms of showing how discrimination, system errors and other problems have lethal consequences in migration and border management, calling attention to the private interests behind these practices, and demonstrating how they contribute to the dystopian direction in which we are headed.

Metaphysical consequences of technological developments

When dealing with comprehensive, complicated and advanced technological products such as artificial intelligence and blockchain, making a technical analysis of these products and identifying their political, social and cultural consequences is a difficult, time-consuming process. Moreover, when dealing with an algorithm that analyses data and draws conclusions, and is considered 'smart', such as in the field of AI, approaches to this special and extraordinary product can be quite metaphysical (Joque, 2022). The most obvious example of this is the view that considers technology to be objective, neutral and scientific. This view hides or obscures the power relations, forms of ownership, interests and exploitation behind technological developments. For example, when an employer is fired, they may blame the boss, human resources or their supervisor, learn about their rights, take legal action or make other efforts to mitigate their situation. However, when AI is involved and the employee is presented with changes in their performance and efficiency relative to colleagues, they would be more likely to accept the situation because the calculation is made by a machine and results are shown in numbers, indicating that it is not personal: they were simply under-serving and lost their job. In another example, in the event of a negative ruling in court or rejection of a visa or asylum application, it is possible to blame the judge, the consul or the border agent involved, attributing the decision to a lack of knowledge or discrimination. However, a decision made by an algorithm is more likely to be met with acceptance on the grounds that it is made by an unbiased agent. People may also be swayed by the way technological products are reported in the media, which usually talk about AI doing this or an algorithm making that, as if they aware of their self-interest, without making any references to the interests of their owners (Angwin et al, 2016).

This can apply to critically minded observers as well, who may state that AI discriminates, makes mistakes, fails or cannot solve a given problem. This can give the impression that there is not a structural problem, but, rather, simply an algorithm that is not well designed or sufficiently developed. In other words, the implication is that these problems could be avoided or structural inequalities would not be manifested if only the algorithm were better trained. For example, in the case of exam grades being marked down by an algorithm in the UK, students protested the algorithm, upon which the government backed down from its decision by blaming the algorithm and cancelling the markdowns. While the students' demands were eventually met, the algorithm in question exposed the class-based discrimination in the British education system and the glaring gap between state and private schools (Kolkman, 2020). Significantly, the episode did not trigger demands for a deeper reform in the education system. Everyone was mostly satisfied when the algorithm was blamed and the markdowns were cancelled; meanwhile, graduates of private schools continue to be admitted to the best colleges and universities, aided by their historical advantages. This development may have helped a small number of highly successful students attending state schools and residing in poor neighbourhoods to get into good universities, but universities already admit a small number of highly successful poor students in the name of diversity and to prevent criticism. The algorithm, in this case, again, served to obscure the powerful groups who owned and used the algorithm.

Metaphysical approaches are not limited to people's treatment of these technological products as neutral and objective. Nor are they limited to these means of production obscuring capitalists who establish private ownership over them and produce values in line with their own interests. The statistical principles underlying these algorithms are another factor in the emergence of these metaphysical approaches (Joque, 2022). Big-data analysis using advanced statistical methods boils down to calculating different probabilities and assigning them various weights. In short, when an algorithm predicts the future or an unknown value, it merely identifies the option with the highest probability. Therefore, this is not about 'trusting' the algorithm. What matters is understanding that any prediction comes with some degree of uncertainty. For example, when there is an 80 per cent chance of x happening, the algorithm predicts x, even though there is also a 20 per cent chance of x not happening. It would be difficult to call this an objective or neutral conclusion. It is simply an exercise in identifying the option with the highest probability.

Statistics generate successful results in the cases of playing Go and chess, for example, without resulting in social damage or discrimination. Artificial intelligence, in these cases, can successfully analyse all the potential moves by its human opponent because the algorithm has played the game thousands of times and analysed thousands of different games. It may also develop interesting combinations or tactics that may not be apparent to human players and so beat human opponents. This sort of analysis can generate successful and effective results in many fields.

When it comes to social issues, however, statistical results can be manipulated in favour of the powerful or to generate mystical results, with probable options being presented as inevitable or real. Thus, when Black people are over-represented in US prisons, an algorithm may decide that a Black suspect is more likely to commit a crime and recommend ruling in favour of a white suspect. Similarly, when the structural factors behind the success of private schools are not considered, an algorithm may decide, looking at existing data, that students attending state schools are more likely to have lower grades. Therefore, it encourages an approach that perceives it as neutral and objective, magnifying the algorithm and obscuring the structural factors (O'Neil, 2017).

This, in turn, can cause serious problems in migration and border management. When a migrant or asylum seeker is categorized and assigned a risk score, along with thousands of other people who are considered to be similar (in terms of nationality, place of origin, income, education and so on), applicants who face a real risk of getting killed or hurt may end up being rejected. Of two people in the same category, one may face a real threat of being killed, whereas the other may not. The algorithm cannot offer certainty in these cases; it only provides a percentage, for example saying that there is an 80 per cent risk. However, 20 per cent is still significant and should not be easily dismissed. Similarly, lie detectors used at border crossing points (which will be discussed in more detail in the following chapters) show how problematic and risky that approach can be. Piloted in many countries, lie detectors using AI algorithms deployed in the form of kiosks have migrant people respond to certain questions. Depending on their answers, gestures and mimics, perspiration, changes in the colour of their skin and levels of anxiety, all measured by sensors, detectors refer them to an official if the algorithm 'works out' that the person is lying (Daniels, 2018). Even if the claimed success rate of 70 per cent is accurate, this is only a percentage, and a 30 per cent chance should not be dismissed. It is not reliable enough to form the basis of a decision regarding a migrant's or asylum seeker's application. These decisions can put people's lives at risk.

Statistics can also be used to analyse human behaviour and discourse, as well as to predict crime. However, with regard to social and political issues with direct implications for actual human beings, as these models simply explain the current data and serve as a mirror to society, we cannot claim their results are an accurate representation of truth or are neutral. Therefore, running pilot AI projects on vulnerable or risky populations may reinforce the risks rather than alleviate them (Milivojevic, 2021).

Yet, the results of the metaphysical approach are visible in migration and border management as well. For example, many people who respect and support migrants' rights, may think, when an application is rejected based on technological products, that it is a scientific truth and choose to remain silent. Many people who would protest if a border agent were to adopt a racist attitude may acquiesce when the process is automated. For migrant people themselves, it may be easier to accept a decision made by an algorithm. This is yet another case of a technological product obscuring serious problems and the political interests behind them. For example, there is a strong reaction in US society to the physical wall Trump has erected on the Mexican border, with Democrats voicing strong opposition, but there is not much difference between the Democrats and the Republicans when it comes to erecting a virtual wall or using the latest technologies, which enjoy much higher rates of acceptance in society (Franco, 2019; Mijente, 2019). This is despite the fact that virtual walls – that is to say sensors, fake phone towers and drones – lead a larger number of migrant people to death and can have more detrimental results than a physical wall.

Are migration-management technologies limited to the field of migration?

Surveillance technologies developed and tested in migration and border management can very well spread to other fields and affect an entire society. Suppose the lie detectors piloted on borders, for example, are successful and meet no significant opposition from society, we should not be surprised to see lie detectors in workplaces, at schools or in police stations in the near future. This, in turn, would mean new mechanisms of oppression in society at large. An employee who undertakes undercover union work at a workplace, for example, may admit out of fear that the detector would expose them; or students, employees or people unrelated to an issue at hand may face accusations and fines because of a wrong call by the detector, be fired, or have their reputations ruined.

As we will see throughout this book, almost all the technologies developed in recent years are in active use in the field of migration and border management. These fall under the category of surveillance technologies (Au, 2021) because tracking, monitoring and stopping migrants, ideally before they reach the border, are of utmost importance in this field. Therefore, the products developed end up reinforcing surveillance.

It is notable that the deployments of these technological products are often referred to as pilots. This is because they are not yet sufficiently developed; the practice is to try them on migrants first, or, in other words, to use migrant people as guinea pigs in testing new technological products (Molnar, 2021a). And we are not in a position to comprehensively understand the risks and troubles faced by people used as human subjects in these experiments.

A significant proportion of the major technologies used in migration and border management are also related to war technologies. Drones used to detect migrants can easily be turned into war equipment and used to drop bombs on other countries (Benson, 2015; Csernatori, 2018; Fussell, 2019; Ghaffary, 2019). People watch dancing robots with amusement, but the same robots are patrolling the US-Mexico border as border agents, trying to catch migrants. It is not a stretch to think that we will soon see them wielding weapons on the battlefield. Moreover, it is possible to use tools, such as sensors and satellites, to detect migrants before they even reach the border, equally utilized to monitor the territories of neighbouring countries for intelligence-gathering purposes. Therefore, as the militarization of borders gains pace with the latest technologies, it is obvious that this would create additional problems in maintaining peace at regional and global levels (Azieki et al, 2021; Bankston, 2021).

Because migrant people have limited opportunities to exercise their freedom of expression and defend their rights, there is little social oversight on these issues, especially in the many contexts where anti-immigrant views are popular within political and societal discourse. Many military and technology companies, along with military and defence bureaucracies, develop these technologies with the aid of academic research groups (Larsson, 2020). Analysis of these networks of relationships, which have gained an oligarchic nature, are important components of this research.

In this book, I will analyse how the latest technologies are used in migration and border management, who is driving them, and their potential consequences. Moreover, I will explain how these fit with the overall trajectory of capitalism and the connection to contemporary capitalism. I believe this is important to provide context and, in part, an explanation for the drivers of border surveillance, an increasingly opposed field in research and NGOs (Albahari, 2015; Akkerman, 2016). It is argued that migrants and asylum seekers are denied their basic rights and used as guinea pigs in trials of new technologies. A case in point is when, in 2022, refugees thrown in semi-prisons – 'high-tech closed refugee camps' – on three Greek islands in the Aegean were monitored using cutting-edge technologies, which effectively turned these camps into experiments in surveillance technology, with the refugees being forced to serve as human subjects (Molnar, 2021b).

Use of surveillance technologies in migration and border management is not limited to just a few countries. Indeed, there appears to be a consensus among advanced capitalist countries. It is not just the US and the UK; the EU and respective EU member states, Australia and Canada have similar policies (Hall, 2017; Keung, 2017; Molnar and Gill, 2018; Akhmetova and Harris, 2021; Dumbrava, 2021). Migrants and refugees are not welcome in Japan, Russia or China. Preventing migrants and militarizing borders are popular not only in the Global North but in the countries of the South as well. Examples include Turkey, Mexico and Bangladesh. Running through this general anti-migrant sentiment in policy, there is a common preference for smart borders and surveillance of migrants. Producers of these technologies are a small number of tech and military companies who market their solutions to the entire world (Akkerman, 2016; Privacy International, 2016, 2021).

Selection, monitoring and control of migrants

Given this state of affairs, it is not enough to observe that these technologies are used on migrants. On their own, human-rights advocacy and defending the rights of migrants are not sufficient either. These rights are already recognized by the European Union and its member states, along with many other countries. Still, the problem is that they have been denied to certain categories of people in recent years. Exclusive emphasis on rights gives the impression that the problem might disappear if only current administrations were replaced. Therefore, these developments should be considered together with the overall trajectory of capitalism, the increasing importance of surveillance technologies for capitalism, and a small number of Silicon Valley companies gaining monopoly status, deepening their ties to and sharing common interests with the military corporations and financial institutions. This would make it possible to understand why migrants are used as guinea pigs, what criteria states use in selecting migrants – opening their doors to some and excluding others, how they outsource this process of selection, and why the risk of war would increase, domestic democratic institutions would be weakened, and societies would face more surveillance if these pilot programmes were successful. Therefore, I will be frequently referencing the structure of capitalism and the relatively recent concept of surveillance capitalism.

Associating developments in migration and border management with the overall trajectory of capitalism will make it possible to focus on the process of how capitalist states select migrants. Answers to questions such as which migrants are stopped and how those allowed into the country are monitored and denied some of their rights while participating in the labour force, will show that this is not just a matter of migration management. Migration management has implications for the labour market with regard to who will participate and to what extent, which sectors need additional workers, and what mechanisms are used to lower wages. In short, management of migrant flows, efficient distribution of the migrant labour force, and selection of migrants who fit these criteria are becoming more important in interventions by advanced capitalist states at the global level to shape and relocate the working class and achieve higher levels of surplus-value extraction. The latest technologies, and new data analysis methods in particular, make it possible to do this faster and much more efficiently.

This book makes the case that these applications of technology offer enhanced opportunities for selecting and steering migrants, as well as for piloting and developing technological products further. In particular, it calls attention to the important role played by data gathered from migrants in training algorithms and urges us to question the process by which UN agencies and humanitarian aid organizations are turning into data suppliers for Big Tech. It is interesting to look at examples that show how technological solutions supposedly developed for the financial and social integration of migrants, and considered pro-migrant, are, in fact, associated with smart border applications that may lead migrants to their deaths. For those interested in analyses of capitalism, understanding the experiments run on migrants would be useful even if migration is not a focus of study.

The term 'migrant' in this book does not refer only to refugees, asylum seekers and irregular migrants who cross borders without permission. I will be using examples mentioning these categories in relevant sections, but many of the technological applications discussed in this book also apply to legal migrants who have residence and work permits. This might apply to people who have work visas, employer/entrepreneur visas, student visas or family reunion visas. Reframing in terms of mobility, rather than migration, technologies also apply to tourism, business trips and short-term travel. Take, for example, the diverse range of people who apply for a Schengen visa to visit Paris for a week. They share all sorts of personal details about themselves for this application. They respond to tens of questions about their level of education, occupation, bank accounts, personal lives and family members; they are asked to submit lengthy documents and are then photographed and fingerprinted in visa processing centres. On top of that, they are asked to pay a large sum of money. We entrust all these personal and sensitive data, including biometric data that plays an important role in identifying individuals, to consulates for a short trip. In most cases, such data is provided to a contractor, which may or may not follow good practices regarding cybersecurity and accountability. As a result, many countries have access to all sorts of personal data on millions of foreign nationals who travel for tourism or business purposes. Leaving aside the risk that this data could be hacked, we do not know for what purposes it will be analysed and what sorts of actions will be taken regarding foreign nationals who volunteer such data. From this perspective, this book should speak to a wide audience.

The book consists of four main. Chapter 1, Migration and (Surveillance) Capitalism, examines how and why tech companies invest in the field of migration and border management along with military companies and finance firms and why new surveillance technologies are tested on migrant people in the context of the general characteristics of contemporary capitalism. More specifically, it focuses on the theory of surveillance capitalism and argues that this theory is not limited to the business models of a few Silicon Valley companies, but represents an intrinsic and structural characteristic of contemporary capitalism. After presenting this general theoretical framework, the chapter introduces the consequences of technological products used in the field of migration and border management discussed in the following chapters, along with examples.

Chapter 2 deals with migration and (big) data. It discusses the results of analyses of data collected from migrants using various means, the relationship to surveillance, the security of migrants' personal data, and relevant ethical principles. This chapter will focus on how the data obtained from migrants is treated, whether collected from social media and smartphones or satellites, drones and sensors.

Chapter 3 discusses migration and smart borders. Smart border applications are based on the analysis of data collected from migrants. Technology makes it possible to collect and analyse this data before migrants arrive at the border and allows states to select and let in 'desirable' migrants, while keeping the rest away. The chapter also assesses tech firms and military companies producing these technologies, and calls attention to their ties to security-defence bureaucracy and academic institutions, their power in shaping the agenda and policies regarding migration and border management, and how they test new technologies.

Chapter 4 deals with digital identity and migration. Digital identity, or the digitization of migrants' identities, is usually achieved through biometric and blockchain technologies. Finance companies and mobile phone operators, in particular, have significant investment interests in this field. The policy aim here is to support the financial integration of migrants using the data collected, while this often contributes to techniques of surveillance.

Previous studies in the literature do not consider the relationship between digital identity and smart borders, treating them as separate subjects and fields. However, the two are connected because smart borders are based on the gathering and analysis of big data, while digital identity is about where, how and by whom this data is kept and analysed. The two also complement one another in terms of tracking, selecting and controlling migrants, as well as monitoring and integrating those who are able to cross the border into their new society. Therefore, one of the aims of this book is to bridge these fields, which are seen to be separate or disconnected, and show that they are in fact different aspects of capitalism's attitude toward the migrant labour force.

In academia, there are different approaches to these topics. Serious concerns and ethical objections are raised regarding big data. Still, there is lingering optimism that data analysis could also be employed to benefit migrants, which would solve problems quickly if only the necessary steps were taken. Smart borders, however, are mostly criticized and viewed as a category of application that disregards migrants' rights. Digital identity, like big data, has both supporters and critics, and many studies focus on how to use digital identity more efficiently and to prevent potential harm.

This book argues that the agenda is set by big corporations and advanced/core capitalist states in all of these main topics, and a similar political approach is adopted to focus on the selection and exploitation of the migrant labour force, ranging from the rejection of some migrants to other migrants' integration and participation in the labour force. This approach is put into practice through programmes that are globally accepted and implemented. In countries that are subject to the most intense migration flows, such as the US, Canada, the UK, Bangladesh, the Philippines, Thailand, Turkey, Libya, Jordan, Brazil and Colombia, these programmes, which protect the interests of the same 'donor countries', are implemented through UN agencies and humanitarian aid organizations. Testing of technologies is not limited to AI algorithms or war technologies either, with, for example, blockchain technology being tested and developed further using digital identities. Because UN agencies and humanitarian organizations do not have the technical infrastructure to develop and deploy these technologies, Big Tech companies enter the picture, and we do not have comprehensively knowledge of what they do with the data they collect.

Using a range of case studies, this book underlines and explains how cutting-edge technologies are actively used in putting this approach into practice, managing millions of people on the move; how data analysis is used to decide who is permitted to migrate; and how migrants are monitored and steered once they cross the border. Thanks to these technologies, a small number of monopoly companies (in technology, finance, military and telecommunications) and advanced capitalist countries (known as 'donors' in humanitarian emergencies that fund various projects) are implementing their agendas throughout the world. The development and testing of surveillance technologies, and their future deployment in all societies if found to be successful in their aims, provides important clues as to how capitalism spreads and takes root. This is the primary focus of this book.

1

Migration and (Surveillance) Capitalism

There is a large body of literature on the relationship between capitalism and migration. Structuralist, Marxist and Marxist-influenced theories examine migration from different perspectives, but mainly in terms of exploitation and relations of dependency; analyses that were particularly popular in the 1970s but remain relevant. There are many studies on how capitalism uses migrants as a labour force to increase surplus-value extraction and to lower wages, and on the tactics used by labour unions to organize migrants in different countries. Similarly, many studies deal with the consequences of migration, both for origin and destination countries (from 'developing' countries to 'developed' capitalist countries), the effects of remittances, the issue of brain drain and continuing colonial ties and racism. There is also an extensive number of studies that adopt a gender approach, for example the migration of women – particularly from countries such as the Philippines and Ethiopia – to work in domestic services, conditions of modern slavery, and the supply chain of care for children and the elderly at the global level (such as scenarios where a Filipino mother looks after the child of a middle-class German mother, whereas the Filipino mother's child is looked after by another woman in her country of origin). Moreover, some studies examine the precarious working conditions faced by migrants to capitalist countries, and migrations among the emerging and more economically dependent countries called the Global South. Overall, these studies analyse the approach adopted by capitalism to migration – of labour, in particular – in different periods and countries (Zelinsky, 1971; Amin, 1974; Gilroy, 1993; Escobar, 1995; Anderson, 2000; Arango, 2000; Bigo, 2002; Creswell, 2006; Castles and Wise, 2007; Raghuram, 2009; Castles, 2010; Gardner, 2011; Cabanes and Acedera, 2012; Browne, 2015; Massey, 2015; Francisco, 2018).

In this chapter, I will make use of the accumulated findings and analyses of this literature, but I will not limit myself to the consequences of capitalist policies in the field of migration; I will also argue, making references to the surveillance technologies developed, that migration and border management policies and practices provide valuable data on the internal evolution and trajectory of capitalism.

Field studies on the use of surveillance technologies in migration and border management focus on the private interests and companies that are behind these investments (Akkerman, 2016; Privacy International, 2016, 2020, 2021, 2022; Molnar, 2019, 2021a, 2021b) and analyse the traumas and violations of rights faced by migrant people, but there is not sufficient emphasis on their relationship with the overall trajectory of global capitalism. A large number of companies and states invest in these fields, but we need further studies and discussion on the main motivations besides profit maximization, such as how alliances are formed in these fields, which actors set the agenda, and how capitalists in many different industries utilize these outcomes. For example, many reports examine the profits made and surveillance technologies developed by Airbus and Thales – as weapons producers – in the field of border management. However, the benefits of these applications accrue not only to their developers but also to Zara and H&M as ready-made-garment companies, as well as to automotive firms Renault and Ford, among others. Military and defence industries and tech companies are not the only ones that profit from surveillance technologies; universities/academia and financial firms also benefit. This chapter discusses the reasons why, and analyses the issue of migration and capitalism from a wider perspective.

Why do we need to discuss capitalism?

There are two main reasons to include discussion of capitalism when examining the surveillance technologies developed in the field of migration and border management. The first, as mentioned towards the end of the Introduction, is the global character acquired by the technologies and 'measures' used to implement these policies. For example, we can see the same political approaches and surveillance technologies deployed in the US, Canada, the EU and its member states, the UK, Israel and Australia. The same can also be observed in countries such as Turkey, Mexico, Saudi Arabia, Bangladesh and Colombia, which can be described as periphery or semi-periphery countries, underlining their level of dependency to the core/ advanced capitalist countries. Therefore, similar policies and technologies are implemented at the global level on issues such as analysing migrants, legal or illegal, before they arrive at the border and deciding whom to allow entry; using technological products to forcefully and violently stop those trying to enter without permission and sending them to places they would not like to go; and monitoring migrants whose entry into a country was permitted and deciding whether they should participate in the labour force, enjoy full rights and have access to banking and social services. All of this is made possible by the ability of core capitalist countries to lead UN agencies and humanitarian aid organizations – which they fund as 'donors' – to take similar action throughout the world, in the fields of migration, border management and humanitarian aid, and the fact that many of the companies that develop and implement these technologies are global tech monopolies.

The second reason, briefly mentioned above and to be discussed in more detail in the following, is that many capitalist actors, institutions and sectors benefit directly or indirectly from the technologies used in migration and border management. Many organizations that reproduce capitalist relations in daily life benefit from this process, including states, academia, companies in different sectors (not just finance, technology and defence industries, but many service and manufacturing industries as well), international organizations (such as the UN, the World Bank, the World Economic Forum) and NGOs. Therefore, evaluating the overall trajectory of capitalism is a must for understanding why surveillance technologies are piloted and implemented globally in the field of migration and border management, and how they are related to class identity, class inequalities and class struggle within individual countries and at the global level.

Critical approaches to migration studies and the use of cutting-edge technologies in migration and border management have created rich literature. Studies on forced migration and mixed-motive migration focus on the effects of military occupations and conflicts. Moreover, the global challenges of the climate crisis, poverty and inequality and technological transformation are examined on this basis. What I would like to underline here is that we should discuss the structural consequences of global capitalism in all these fields, not only of migration, but also of the climate crisis, (neo)colonial relations, conflicts and occupations, racism, poverty and inequality. Analysis would be incomplete without an evaluation of capitalist relations of production and conflicts of interest. For example, several factors, from racism to colonial relations, arguably play a role in cases of using drones to detect migrants in the Mediterranean and then pushing them back, and accepting asylum applications from some countries while summarily rejecting applications from others. These observations may be accurate, but we also must deal with the causes and roots of racism and colonial relations (Broeders, 2007; Hampshire and Broeders, 2010; d'Appolonia, 2012; De Genova, 2013; Albahari, 2015; De Leon, 2015; Csernatori, 2018).

Without a structural approach, we may bias the approach toward targeting authoritarian administrations, dictators, right-wing parties in government in the US and EU member states, bureaucrats with racist ideologies, or right-wing populist leaders trying to ride the popular discontent. Many reports and academic texts that defend the rights of migrants make analyses discussing the responsibility of these groups and criticize them. They do have a point because, if migrants are being left to die on the border, we cannot deny the responsibility of bureaucrats, political leaders and incumbent parties. However, assuming that problems would be solved if only these people were replaced can prevent reaching the right conclusions. For example, the wall Trump has built on the Mexican border of the US is criticized by the Democratic Party, but the two parties agree on using cutting-edge technology to raise virtual walls on the border. The argument is that the use of new technology on the border would have functions such as making objective decisions and preventing border officials who might harbour racist ideologies, but migrants are then forced to attempt crossings in the most dangerous parts of deserts and seas because of sensors, drones and satellite images that can see beyond the border, and these virtual walls result in more migrant deaths. In any case, the Biden administration that succeeded Trump has so far continued the same basic arguments and policies regarding border and migration management (Mijente, 2019; Akhmetova and Harris, 2021; Aizeki and Shah, 2022). The same can be said about EU member states (Germany, France, Denmark, Italy, Spain, Greece and Hungary, in particular), Canada, the UK and Australia, where parties with different political views are in power. Therefore, it would not do to blame the whole thing on Trump, Johnson, Erdoğan, the Greek government, the Director General of Frontex or the European Commission. In many of these countries, opposition parties are not proposing alternative policies in migration and border management either. Given the wide scope involved, with governments and relevant political institutions at the global level and companies operating and making profits in this field, I believe structural characteristics and evolution of the capitalist system should feature more prominently in analyses of the use of surveillance technologies in migration and border management.

The free movement of labour

If we are to look more closely at the overall trajectory of capitalism, we must start by adopting a class-oriented perspective. Of course, it is not possible to provide a comprehensive look at the main class dynamics of capitalism within the confines of this book. However, just as Marx began his analysis of capitalism by analysing the smallest building block 'commodity' and then offered a gradually expanding perspective on surplus-value extraction, relations of production and private ownership of the means of production, I believe it would be useful to begin the discussion with the class position of migrants.

Studies on migration and refugees classify migrants in different ways, such as skilled or unskilled labour, capital owners, students; regular and irregular migration, domestic and international migration, migration due to climate crisis, forced migration; asylum seekers, refugees and those granted temporary protection status. Studies also focus on different time periods, regions or countries. This categorization is not specific to academia. In international law, developed after the Second World War, UN treaties, bilateral agreements between countries, and the domestic laws of individual countries define different types of migration and come up with specific criteria for each.

With changes in the field of migration, however, new developments have become a topic of discussion, from the role of companies and NGOs in migration and border management to diverse causes of migration. For example, one of the most recent debates revolves around how the asylum and refugee laws developed from the 1951 Refugee Convention is no longer sufficient to explain contemporary migration flows, as those texts were based on the experiences of individuals facing particular political persecution and thus offer an insufficient legal framework to deal with the hundreds of thousands or even millions of people fleeing wars today, and new approaches should be developed. International initiatives like the Global Compact for Migration (United Nations, 2018), UN agencies, donor countries and countries facing forced migration flows propose new legal approaches such as 'temporary protection'. Within academia, the argument is put forward that there is no longer a meaningful distinction between migrants and refugees and all refugees are migrants, and the multiple causes of migration (such as poverty, conflict and climate crisis) are brought together under the term 'mixed migration'.

Notably, many global, international or national corporations have also taken part in migration management and border management as active stakeholders. Migration control and border management were long considered to be a public service; meanwhile, the role of companies was limited to lobbying to influence labour policies. The legal framework regarding whom to allow to cross the borders, and the rights and freedoms of those allowed entry, was created by public authorities. It is true that capitalist states acted to protect the interests of capitalism, but the role of private companies was limited to providing input in the making of policies and supplying products and services during humanitarian crises. In recent years, however, tech companies, military/defence companies, financial firms and mobile phone operators have been working on projects regarding migrants and refugees, and many companies operating in manufacturing and service industries directly benefit from these processes. Their work extends from undertaking social-responsibility projects to entering new markets and reaching new customer segments, and from product development to the supply of big data (discussed in more detail in Chapter 2). Professional associations, alliances and initiatives are being formed in many sectors (finance, telecommunications, technology and textiles, in particular) with the specific aim of undertaking work on migrants and refugees. Alongside states, many global companies also fund UN agencies, academics and humanitarian aid organizations as donors (Mercy Corps, 2016; GSMA, 2018; Mastercard, 2019).

In their capacity as donor countries, advanced capitalist countries (the UK, the US, Germany, the Netherlands, Sweden, Switzerland, Japan and France) already set migration and border management policies. Global companies also take part in agenda-setting and framing as active stakeholders, without limiting themselves with social-responsibility projects or the provision of products and services. If a field with such comprehensive interests and institutionalization also carries out work that affects hundreds of millions of people on almost all continents of the planet, then it is only natural for such work to provide important clues to the overall trajectory of capitalism. For example, the World Food Programme (WFP) alone distributes aid to close to 100 million people, and the UNHCR and International Organization for Migration (IOM) deal with more than 80 million migrants and refugees. Given that the direct beneficiaries of these organizations are in daily contact with many other locals who live in the same places, these programmes affect the lives of hundreds of millions of people.

How, then, can we explain the causes of this interest? It might be useful to recall neoliberal policies to overcome barriers in front of the free flow of goods, services and labour. The concept of globalization, which gained prominence in the 1990s after the collapse of socialist states, was based on the notion of free movement of goods, services, capital and labour, which was also a key feature of the process of European integration. Many international organizations and states thus followed neoliberal policies to remove obstacles to this free movement and carried out comprehensive reforms to this end. For the purposes of this book, what matters are the policies followed regarding migrant labour, within the context of the 'movement of labour'.

From the end of the Second World War to the 1980s, the employment of migrant workers was under the control of states and usually governed by bilateral treaties between countries. One of the best-known examples was the guest-worker programmes Germany signed with countries such as Greece, Turkey and Spain. Similarly, migrant workers from the UK's former colonies in the Caribbean and South Asia were meant to close the labour gap in the advanced capitalist countries. Domestic and international migration have increased significantly due to the free movement of labour, advances in transportation and communication technologies and rapid urbanization at the global level. These, in turn, have reshaped the role and function of the state, in particular the new roles and opportunities that the latest surveillance technologies offer to governments in terms of migration and border management and the new opportunities that global companies – especially those operating in technology, defence, communication and finance – have found as active stakeholders in migration movements. The decline in profit margins in the global economy, ongoing economic crises, increasing competition among capitalist states and companies and long-running or endless wars and conflicts in many regions of the world (Fuchs, 2019, Morozov, 2022), provide a crucial background regarding the importance of surveillance technologies for the trajectory of capitalism.

The role of migration policies and surveillance technologies in shaping the international working class

On the one hand, capitalism needs migration flows to increase labour efficiency, reduce wages and deepen exploitation, and aims to ensure the continuity of migratory movements. On the other hand, when unexpected and rapid mass migration movements take place (for example, when millions of people flee to neighbouring countries from conflicts or natural disasters), migration needs to be managed at the global level. This must be planned in such a way that countries in different levels of capitalist development, with different roles in world economy, that offer different products and services and have different levels of dependency can meet their labour needs with the migrant workforce while preventing entry of migrants not useful or needed by them.

What I have briefly described here is the process by which global capitalism reshapes and repositions the global proletariat. Throughout history, capitalism has made efforts to reshape and steer working classes both at national and international levels. Thanks to this dynamic process, capitalism aims to guarantee the continuity of surplus-value extraction. Of course, this is not a one-way process. The working class, in turn, tries to change the situation and reduce exploitation through labour unions and political organizations. In this conflict, as companies try to suppress the demands of local workers by including the migrant workforce in the process, labour unions try to preserve their bargaining power by organizing migrants, especially in core countries (Pulignano et al, 2015; Korkmaz, 2018).

In this very process, surveillance technologies provide capitalism with immense power to shape the global working class. At a time when large masses of people move quickly, these technologies make it possible to identify migrants before they even arrive at the border, analyse their data, decide on whether to allow them entry or not, and closely monitor them if they are allowed into the country. This process of selection and the interests of capitalism in shaping the labour market constitute the essence of the issue.

Therefore, discussions about the legal categories of migration, mentioned at the beginning of this chapter, become *de facto* pointless. The legal framework for these categories is ignored, becomes inoperable or is openly rejected. The main determinants are the capitalist interests, whether they need certain migrant communities with average skills and education or not.

Surveillance technologies jointly developed by military companies and technology firms allow new members to join the international working class from places where the reach of global capitalism is relatively weak (such as Afghanistan, Syria and Yemen), while making it possible, based on the global division of labour, to identify who would be most useful in which regions and for which companies and sectors, and to steer migrants accordingly. Within this framework, migrants or refugees can move until they arrive at the places designated for them, but when they try to go further or work in a different field, they are stopped by states, which, equipped with surveillance technologies, continuously monitor them, limit their movements and measure their efficiency.

Adopting a market-oriented perspective, companies from many sectors invest in this field. Military and security firms develop products to stop undesired arrivals, trying new military products on migrants, whereas banks and mobile phone operators develop new products and services to integrate migrants into the financial sector, reaching new markets and potential customers in the process. Technology companies, in addition to offering products in cooperation with other sectors (such as military, finance and communication), get the opportunity to access data from hundreds of millions of people who are usually not protected by privacy laws, which allows them to train their algorithms. Migrants, on the other hand, unable to go beyond the limits imposed on them, start working for the global supply chains of sectors such as textiles, food, automotive and construction.

As most migrants are labourers, big-data analysis makes it possible to measure and calculate the potential value created by such workers. As discussed in the following chapters, the main goal in efforts for digital identities or digitization of identities is not limited to providing people with an identity and facilitating their access to basic services; arguably, the main motivation is to meet 'know your customer' requirements, which are necessary for opening a bank account or purchasing a SIM card (Slavin, 2021). This allows holders of digital identities to generate data quickly and become cogs in the machinery of capitalism as clients, consumers, employees or suppliers.

So, people may be migrating for diverse reasons, and there might be legal categories corresponding to each in order to manage them better, but these are fast becoming pointless, and an agenda based on increasing the profit margins and market shares of capitalist companies is being imposed in the fields of migration, border management and humanitarian assistance. Initiatives for 'banking the unbanked' – an agenda that companies imposed on UN agencies and humanitarian aid bodies – might be the most obvious example of corporate agenda-setting (Mastercard, 2019). The outrageous description of forced migrants as 'the unbanked' requires no explanation. Technological advances and products involving blockchain, AI and data analysis, as well as debates around the themes of innovation and development, make sense within the framework of this agenda and logic (Coppi et al, 2021).

Because, in the absence of structural analysis, it is not possible to see the actors and actions of capitalism with such clarity, academic debates on the causes of 'mixed migration' and on questions such as, 'Are refugees migrants? Has this distinction disappeared? Should refugees stay in camps or cities?' are simply insufficient. The goal is for the migrants to join the labour force of the host country as soon as possible. To this end, UN agencies and humanitarian aid organizations provide vocational training to migrants using the millions of dollars they receive from donors, and if some of the migrants have the potential to establish companies, enter the global supply chain and employ other migrants, they are given entrepreneurship training, mentorship and access to accelerator programmes. This sort of work is popularly named the 'humanitarian-development nexus' (Toyama, 2011; Mayer-Schönberger and Ramme, 2018).

The main goal of these campaigns, guided by donor funds and companies, is to make sure that people migrate to the most suitable countries based on the needs of global capitalism, and start working and producing as soon as possible. This might mean the migration of doctors and coders from some countries, and textile, construction and agricultural workers from others. Agencies and NGOs that greet them on the border look forward to the migrants opening bank accounts and purchasing SIM cards for their phones, and reporting it as a success story to donors when migrants are placed in jobs.

Therefore, the answer to whether refugees should live in camps or among the local population depends on the position of the country in question in the capitalist division of labour. In countries such as Turkey, which has advanced supply chains and many small-scale workshops, refugees staying in camps is not beneficial to capitalism: it does not further the interests of local or international capital. Therefore, camps are closed and refugees start living among the local population, working in small workshops without work permits (Korkmaz, 2017). In Jordan, on the other hand, it is preferable for refugees to stay in camps and be shuttled back and forth, on buses funded by UN agencies, to work in textile factories in 'special industrial centres' established in the middle of the desert. In both cases, issues such as preparing the migrants to participate in the workforce, providing them with vocational and language training, obtaining permits for the firms that employ them, and paying the wages and social-security contributions of the migrants when needed are designed in line with the interests of international corporations and their local suppliers, not with the best interests of the migrant people in mind.

Selecting workers and limits set by capitalist interests

An AI algorithm used by various NGOs in the US offers an interesting example of employing migrant people. Because of public discontent with the concentration of refugees in a few countries, many countries are using resettlement programmes to accept refugees from other countries selectively. In this context, the US accepts a certain number of refugees from countries such as Turkey and Jordan each year. The main criterion of success for NGOs, who facilitate the resettlement process; indeed, the criterion that ensures the continuity of their funding, is that the refugee finds a job within 90 days and stops receiving social aid. This motivation means it is not acceptable for refugees who do not have sufficient English skills to spend too much time in language schools. Therefore, many of them are employed in slaughterhouses that do not require a high level of English, despite common associations with bad working conditions. The welfare, preferences and traumas of refugee people do not really matter; they just have to get a job as soon as possible, even if that means not learning the language for many years and working a job that does not require developing relevant skills.

Turning people into labourers quickly, without regard to the causes of their migration, is also seen in what many consider to be positive examples of refugee policy. When Germany, for instance, adopted such a policy in 2015 and admitted close to 1 million Syrian refugees, it had a particular labour gap in the metals industry and collective bargaining talks with labour unions were just about to begin. The sudden injection of hundreds of thousands of people into the workforce had a decisive effect on wages and bargaining power. At the time, German industrialists observed that the new arrivals had paid about €10,000 each to smugglers, and anyone who could pay this sum had to be a qualified person, so all they had to do was teach German to the new arrivals (Korkmaz, 2017).

A similar approach can also be observed regarding refugees from Ukraine. Europe was commended for opening its doors to Ukrainian refugees, but this behaviour was compared with policies of closing the door to people from other regions and forcefully sending them back, and the stark contrast was attributed to various factors, including racism. What was notable in this case was that people who were living 'normal' lives before the Russian invasion hastily left their country after the occupation to save their lives, leaving their homes, careers, relatives and loved ones behind. Though they were severely traumatized, they were rapidly placed in schools and jobs, and found themselves, all of a sudden, working or studying in another country. Television footage showed these refugees being applauded and showered with flowers as they went to their new schools and workplaces, and this rapid and forced transformation was presented as a success story to be emulated. However, it is not normal to have people who have suddenly lost or left behind their homes, loved ones, cities and jobs start working or studying immediately without giving them a chance to take a breath, receive psychological support, acclimatize and decide what to do next. However, as argued earlier, capitalism is not interested in dealing with people's problems, regardless of their backgrounds or causes of migration, or whether they are migrants or refugees. They are new additions to the workforce and must start working as soon as possible.[1]

Shuvashish Thapa (2022) describes the difficulties Nepali students face as they pursue studies in other countries, which provides interesting data on workforce selection. If you are a Nepali citizen who agrees to work in low-paying jobs, you can travel to India or Gulf states without a visa. If, on the other hand, you manage to be admitted to an overseas university, you must spend at least two months collecting the tens of documents required by these countries, deal with a large number of bureaucratic agencies and be prepared to pay a hefty sum given the economic conditions in Nepal. This, in fact, has to do with the role assigned to Nepal in capitalism and its socio-economic reality as a poor and dependent country, as capitalism designates Nepali people as unskilled laborers.

The UK and Germany, for example, grant work permits to tens of thousands of professionals from Turkey, including software engineers, physicians and academics, while rejecting applications for occupations considered to be lower-skilled. They import domestic workers from the Philippines and agricultural workers from Romania, and they admit Ukrainian refugees while rejecting those from Afghanistan or Yemen. However, I would like to emphasize that the main issue behind these attitudes is the effort by capitalism to shape and categorize the working class.

The issue of migrants facing a hostile environment is a frequent topic of discussion in the UK, and the harsh attitude of the UK to asylum seekers and irregular migrants is often criticized. While the policies create a hostile environment, before Brexit, Turkish citizens, for example, were able, thanks to a visa based on the Ankara Agreement, to establish a company in the UK by simply presenting a brief business plan, without any capital requirements, and tens of thousands of families made use of this visa. From 2022 onwards, new college graduates are given two years of residence and work permits to stay and find jobs in the UK. Similarly, academics, programmers, physicians, engineers and other skilled workers are given various advantages through different categories of visa, including the global talent visa and start-up visa. For example, people are given the right to work without a sponsor, stay unemployed or establish their own companies, and obtain a permanent residence permit within three years. The different attitudes toward different migrant groups can be attributed to the desires and needs of finance capital and the rentier capitalism dominant in the UK. Firms operating in the agriculture and hospitality industries, meanwhile, have a relatively weak position in the UK economy, and their complaints about not being able to find enough workers to hire are dismissed. Here, the goal is not to stop migration altogether; on the contrary, incentives are even provided to ensure the continuity of migration flows in fields deemed necessary by British capitalism.

That capitalist approach, to recognize migrants solely as a workforce, is also clear in the Readmission Agreement signed between Turkey and the EU. Under this agreement, Turkey retains Syrian, Afghan, Iranian and Iraqi refugees who take asylum in Turkey within its borders, prevents them from travelling on to Europe, readmits those captured while trying to enter Greece and, in return, receives financial support from the EU. The policies of forced relocation and compelling migrants to stay in a certain country in return for financial gain are criticized by academics within the framework of human rights and the right to asylum, in particular, and many NGOs and journalists oppose these policies.

The issue is no longer limited to anti-immigration, violations of human rights, or the use of refugees as a political bargaining chip by the Erdoğan government in Turkey, which receives regular financial support from the EU and yet silences all political criticism of its actions. There is no doubt that these are important outcomes. However, the model exemplified by this agreement is a very profitable one for European and Turkish capitalists and is part of a process of reshaping the working classes in both the EU and Turkey, lowering wages and increasing production, that is to say, deepening relations of exploitation and pacifying the working class.

Physical walls are being erected, and various technologies (drones, cameras, sensors) are used for monitoring purposes both on the western borders of Turkey with the EU (Greece and Bulgaria) and on its eastern and southern borders with Iran, Iraq and Syria. A significant portion of these investments is funded by the EU. Since 2020, however, Turkey has been following what can best be described as an open-door policy on its border with Iran. The technologies in question are not used to prevent border crossings or register new arrivals, and hundreds of thousands of Afghan, Pakistani and Iranian people cross the border into Turkey every year. Once they cross the border, they immediately go to different cities thanks to existing networks and are quickly employed in sectors such as textiles, leather, construction and agriculture. Because they are not registered, they work as irregular migrants without work permits or social benefits, often in poor working conditions. In addition, there are tens of thousands of migrants from central Asia and Black Sea coastal states who work in domestic services, agriculture, textiles and many other sectors even though they arrive in Turkey on tourist visas and do not have work permits. There are about four million Syrian people in the country, a figure that would be higher if other asylum seekers and irregular migrants were included in calculations. Some of these people arrive in Turkey dreaming of travelling on to Europe but end up remaining in Turkey or start viewing Turkey as a destination country because crossing into Europe has become much more difficult while there are job opportunities in Turkey due to demand for their labour.

Turkey has faced increasing poverty because of the depreciation of the Turkish lira and the economic crisis that has been going on since 2021, despite Turkey's consistent economic output and sustained growth. Given that Turkey has advanced supply chains in sectors such as automotive, textiles, agriculture and food processing, in line with its position in the global economy, and there is widespread manufacturing for global corporations in its sectors, the increasing importance of the migrant workforce makes more sense. Hundreds of thousands of migrants work in supply networks of global apparel companies such as Zara, H&M, M&S, Primark and Next, and other industries such as chemicals, agriculture and metal, without work permits, for long hours and in return for low wages. Thanks to these arrangements, both Turkish capitalists and the European and US companies they supply increase their profits. This suppresses wages in the country and limits the opportunities for workers in Turkey to unionize and engage in collective bargaining.

Labour unions in Turkey cannot organize local and migrant workers together to oppose the process because migrant workers, whenever they become part of a dispute or demand better working conditions, are immediately confronted by security forces and deported on the grounds that they have entered the country illegally. Many migrants, particularly from Afghanistan and Iran, are not officially registered because they cross borders without permission, but it is possible to surveil them using various sources such as smartphones and social media data. The fact that the police, who look the other way as migrants work informally, can step in and deport them in case of a dispute creates extra pressure on migrants, and, in many cases, migrant workers pressure each other to remain silent. The demands of both Turkish and European capitalists for suitable workforces are thus met, and migrant workers are left alone as long as they keep working within the limits set for them, implicitly under threat of deportation if they dare step out of line.

As we will see, this is not limited to global monopolies. Small and medium enterprises, as well as large companies, in countries hosting migrants benefit from the migrant workforce, which reinforces their position within global supply chains and allows them to increase their profits. Moreover, the middle and upper classes also benefit from cheap and informal migrant labour employed in domestic services, agriculture and service sectors.

At this point, it is fitting to use an example to summarize the discussion and explain the goal of capitalism to reshape, reposition and steer the global working class. As long as they remain in their own countries, potential migrants and refugees in Syria, Afghanistan, Yemen or Turkmenistan and, to some extent, Pakistan, India or Nepal are not able to become part of the international working class, as they do not work on fields or in factories that are part of international supply chains. To become part of the global working class, that is to say, to participate in the production of surplus value, they have to go to Turkey to work in the textile industry, for example, or, if they are living in Nepal or India, to the Gulf countries to work in construction and tourism. Once they take the trip to these countries, they appear on capitalism's radar. They produce value as workers while also becoming clients, consumers and data providers for finance, communication and technology companies.

However, these people must be prevented from leaving Turkey, for example, if this is where they are needed. At this point, the use of surveillance technologies makes it possible to identify and halt people who are not planning to stay in Turkey. In short, a Syrian person should go to Turkey to work for a global company but must be prevented from going to Germany. NGOs, UN agencies and various bureaucratic organizations are making great efforts to ensure that they stay in Turkey and adapt to the labour market, acquire relevant skills and learn the local language there.

Collecting data from citizens of other countries

The issue of shaping the international working class is not limited to irregular migration, forced migration or mass population movements (due to wars, earthquakes, volcanic eruptions, flooding or fires), nor to migration from 'developing' countries to 'advanced' capitalist countries. It also concerns legal migration and migrations among wealthier capitalist countries. Moreover, because of the hunger for data and desire for control, it also applies to tourism and all sorts of mobility in general.

Let's say you are a programmer in Turkey, and you receive a job offer from a company in Ireland, Germany or the UK, or you were admitted onto a master's course at a university on full scholarship. To obtain your residence and work permits, you need to do more than submit a few documents such as your ID, job proposal or admission letter from the company or the university in question, and your diplomas. Under normal conditions, proving your identity and submitting an official document from the organization that makes you a proposal should be sufficient. However, the practice of collecting data from citizens of other countries has reached such a point that you are now required to submit sensitive documents about your whole life. You must collect tens of documents totalling hundreds of pages on your financial situation, bank account movements, educational attainment and marital status, among others. Even that is not sufficient, for you are required to provide your biometric data and have your picture and fingerprints taken.

The procedure is not much different when you want to visit a Schengen area country or the UK as a tourist. For a three-day trip to Paris or Amsterdam, you are asked sensitive data, including about your personal life, in addition to providing biometric data. In addition to submitting all this data, you are required to pay a fee every time you make an application. And, at the end of this process, you may be denied a visa without any explanation whatsoever. It is unclear what happens to your data or where it is kept in the event of approval or rejection. What is more, you do not even submit this data to the consulate or to an official agency of the state in question; you give it to an intermediary company designated by that state. It is unclear how this data is stored and transferred, what level of cybersecurity is implemented and whether other analyses are conducted on the data. Principles such as transparency, disclosure and accountability do not apply. As a result, the UK, the US and Germany, for example, accumulate data on millions of people from countries such as Turkey, China and India.

This process can also be viewed as an intelligence activity (discussed in more detail in Chapter 3). In terms of data analysis, there is a huge amount of data transfer taking place. Automation of visa processes and the use of AI algorithms are fast becoming standard practices (discussed in Chapter 2). What I would like to emphasize in this chapter, however, is that these practices are related to class expectations. People who obtain work permits, that is to say, the right to join the workforce of the country they apply to, submit all their data at the beginning of the process, which makes it easier to monitor them for the duration of their permits. It also makes it possible to identify, track and, if need be, detain and deport those who overstay their work permits, student visas or tourist visas. Another important output is predicting that certain people are applying for a tourist visa with the intention of getting informal work, and then rejecting their applications (calculating the probabilities involved is the job of statistics). This also applies to family reunion visas. Testing the language skills and country-related knowledge of those who arrive to join their spouses, either before their application or after they obtain their visas, in order to see if they meet the minimum requirements, deciphering whether the marriage is only 'on paper' and collecting other information are meant to ensure the newly arriving spouse also joins the workforce as soon as possible.

I am not questioning the sovereignty of states here. States may want to know more about the people who would like to cross their borders and live in their territories, make various assessments, grant certain rights depending on their assessments, and engage in political and economic planning accordingly. What I would like to emphasize here is, first, this selection is made based on class interests; second, surveillance technologies such as biometrics, digital technologies and AI make it possible to run these analyses in a much more detailed manner, before the migrant people even leave their countries; and third, the validity of an applicant's claims (such as a job offer or university admission letter) can be verified with fewer documents, but instead, a lot more data is requested from the applicant. Leaving privacy issues aside, the problem here is that states are collecting all sorts of data about citizens of other countries, from biometric data to information on their personal lives. It is unclear what else they can do with these data because there is no limit. This can be a potentially serious problem both for the rights and demands of people who would like to migrate to or visit other countries and for the security of the countries in which they are citizens.

Chapter 3 discusses how surveillance technologies are tested on migrants. The technologies being tested are not limited to those aimed at preventing people who attempt to cross borders irregularly; they also have military applications that could be deployed on the battlefield. All sorts of new technologies, ranging from those used to check biometric data when you try to enter a country as a tourist, student or worker, to AI lie detectors that have speculative features and are disguised as academic projects, are developed using data from citizens of other countries and tested on citizens of other countries. Colonial and dependency relations are reproduced at the level of individuals, and, if these technologies turn out to be successful, they will be implemented at the level of entire societies, strengthening surveillance capitalism's hold.

Discussing surveillance capitalism in border and migration management

We have explored the use of surveillance technologies in migration management in the context of structural characteristics of capitalism and its goal of shaping the international working class. To build on this, the theory of surveillance capitalism offers a useful perspective when discussing the issue in terms of the evolution and future trajectory of capitalism.

It is important in terms of understanding why new surveillance technologies are used in border and migration management and why so much investment is made in this field because, as discussed in Chapter 3, it will help us understand the main motivation and causes behind the investment in cutting-edge technologies used on, for example, the borders of the US or EU, or in high-tech refugee camps set up on Greek islands. This is because, in some cases at least, it is difficult to justify the amount of money being spent, as it does not make much business sense. Taking this as our starting point, we can discuss the expectations from testing these technologies (AI, blockchain, satellites and drones) on migrants and refugees (Macias, 2019; Molnar, 2021a; Bircan, 2022). We can also assess how the successful development and testing of these technologies can increase the surveillance and oppression of an entire society.

All these developments make it possible to diagnose the political and economic alliances being formed in the fields of border and migration management and humanitarian aid. When we look at the potential scope of these technologies and the sizes of the companies and bureaucracies investing in and reaping benefits from these fields, we can also make projections about the overall trajectory of capitalism – and the consideration of 'surveillance capitalism' is a useful reference point for this interrogation.

Development of surveillance capitalism

Since it involves affecting, managing and directing certain people and communities in a systematic, routine and focused manner, surveillance aims to categorize peoples and societies and to have those categories of people carry out their functions within pre-determined limits (Au, 2021). Even though surveillance had always been a significant subject in the history of class societies, it became a basic criterion in both the workplace and political/public life thanks to the development of capitalism. For the capitalist system, the creation of citizens for the nation-state and maximizing the discipline and surplus value of the labour force, from schools to the army and the workplace, are significant issues.

Surveillance capitalism, distinctly, refers to the business model describing the sources of income of some Silicon Valley tech firms and how they developed monopoly characteristics within a short period of time. The leaps in machine learning and the developments of digital technologies (smartphones, apps, the internet of things and so on) create enormous data sets that can be stored in cloud systems and provide great opportunities for companies and states. Thus, the status quo can be subjected to a comprehensive analysis using real-time, cumulative data, and the future can be forecasted, in the ordinary course of the matter, through the analysis of existing data.

Technology companies such as Google and Meta have access to the data of billions of people thanks to the platforms they provide (search engines and social networks), and by analysing such data through strong algorithms, are able to present user-specific advertising. In this business model, the most significant promise to the advertiser is the ability to forecast the future. Accordingly, these platforms present advertisements to the precise people interested in the products or services offered by the advertiser. Therefore, it is only natural that this business model will grow by focusing on the collection of more data, by analysing data in a stronger fashion, and by generating more up-to-the-point forecasts, which require even more thorough surveillance.

Once this business model proves its success, up-to-the-point forecasting alone will not be enough. To secure the revenue stream, clearing away the uncertainty in forecasts is of vital importance, and, accordingly, manipulating the user who observes the advertisement will eliminate this 'problem' (Zuboff, 2019). There have been many critiques against the testing of surveillance technologies, pointing to the manipulation potential of companies working on nudging, persuasion science, profiling, micro-targeting, data-mining, behavioural psychology, social engineering and subliminal techniques, and warnings that vulnerable groups (such as children, refugees, people with disabilities) can be harmed both psychologically and physically (Au, 2021).

States and political bodies very quickly recognized the power of analysis, estimation and manipulation, and have used them as an effective tool in the detection and manipulation of floating voters – an issue discussed extensively during political campaigns in the US (Trump) and UK (Brexit). Furthermore, thanks to their historical accumulation of knowledge and experiences, the security, military and intelligence institutions have also identified associated opportunities, upon which they have begun to act in collaboration with tech companies, and have become the customers of miscellaneous surveillance technologies.

Surveillance capitalism seeks to turn as many activities as possible, live or not, into data to show tailor-made posts to users so that they spend more time on particular platforms and are personally targeted with messaging and advertising (Sadowski, 2019); to involve all economic sectors and other areas, such as the education and social, cultural and personal lives of people.

In summary, when the products developed by these tech companies and the business model they have discovered are merged with surveillance – one of the main principles of capitalism – a structure is created that exceeds its own boundaries and encapsulates the entire system. Like financial firms, these tech companies become global monopolies that could cover all sectors of the economy and appeal to all classes in the population.

Surveillance capitalism and an oligarchic structure

That surveillance capitalism has stopped being a mere business model and became an intrinsic and structural feature of the capitalist system does not stem only from the fact that surveillance is essential for capitalism. I would like to call attention to an alliance among the main actors and sectors of capitalism, which becomes obvious when we look at the developments in border and migration management and humanitarian crises. This alliance plays an active role both in setting the security-oriented policies that lead to the militarization of migration and border management and in developing and testing the required surveillance technologies. This process, in turn, means that the surveillance capitalism business model developed by tech companies is now more than a business model.

The main actors in this alliance include military firms (for example, Elbit, Thales, Airbus), tech companies (such as Palantir, Microsoft, robotics and data-mining companies, social media platforms), telecoms companies (particularly smartphone operators), financial firms (banks and companies like Visa and Mastercard), the security/military bureaucracies (such as Frontex, the UK Home Office, US Customs and Border Protection) and academies (various EU Horizon and UKRI projects) of advanced capitalist countries. The following reasons can be given to describe them as an oligarchic structure.

First, these are among the most powerful and influential actors of global capitalism. Many weapons manufacturers, tech companies, telecoms companies and financial firms already enjoy monopoly status in their sectors; they are few in number and operate at the global level. They are also known to have close ties to the security and military bureaucracies in advanced capitalist countries. This is not limited to weapons manufacturers. Amazon, for example, is the largest defence contractor in the US as it provides cloud services to the Pentagon, and corporations such as IBM, Microsoft and Palantir compete with one another for a large number of defence industry contracts. These companies offer cloud, robotics and AI products and services in line with the needs of the military and security bureaucracy. Moreover, the relationship between these companies and the military/security is also reproduced at the individual level. For example, retired Pentagon generals sit on the boards of many tech firms, and these organizations often engage in personnel exchanges with one another (Akkerman, 2016; Privacy International, 2016, 2019, 2021, 2022; Delcan, 2019; Au, 2021).

Companies in telecommunications and finance industries, similar to weapons manufacturers, can be described as technology companies based on the products they develop, and, because they are technology-oriented, they collaborate with other technology companies. Mobile phone operators and financial firms play an important role both in monitoring migrants and turning them into workers and customers in the countries they migrate to. Coordination between these sectors makes uninterrupted surveillance of migration flows possible, starting from the origin countries and continuing all the way to the end of the migration process. All stages, from preventing some would-be migrants, allowing the actual migration of some, and their participation in the workforce of the host country, can be controlled and directed. Analysis of data collected in this process is used in deciding how to allocate the arriving migrants.

There are special working groups and professional associations focusing on refugees and migration flows in all of these sectors. For example, Microsoft, MasterCard and the GSMA (an association of mobile operators worldwide) undertake work on digital identities and humanitarian aid. Moreover, there are working groups in the EU, UK and US that bring various sectors together with state bureaucracies and provide consultancy and lobbying services. (These are discussed further in Chapters 3 and 4, with examples.) The point is that all of this work in different fields and sectors is coordinated by the military and security bureaucracies of advanced capitalist countries – the donor countries – that deal with migration and border management.

As the examples of Frontex in the EU and the Home Office in the UK show, all work on migration and border management is designed centrally, from setting the policies to be followed to identifying the technological products needed and the tenders to be put up. These processes feed into one another. Problems defined from the security-oriented perspective of the military/security bureaucracy and the lobbying work and proposals of companies in the above-mentioned sectors feed one another. For example, surveillance technology products tested by the Israeli security company Elbit on Palestinian people are then marketed to Frontex and the UK thanks to this 'experience'.

The leading role of the military and security bureaucracies in this process, which have almost exclusive jurisdiction when it comes to dealing with the issue of migration, explains how the security-oriented approach to migration has become the dominant one, if not the only one, as has been discussed in the literature for many years, and how classifying migration and migrants as threats along with terrorism and illegal drugs plays an important role in detecting, monitoring and steering migration movements. This also results in many academic projects being designed to develop the technologies to be discussed in Chapter 3, alongside military and tech companies, under the disguise of conducting research and development work or running pilot projects, and, for example, piloting new technologies under the disguise of developing a 'lie detector'.

In short, the experience of military, security and intelligence bureaucracies in the field of surveillance is brought together with the 'solutions' of tech companies pertaining to the business model of surveillance capitalism, which, thanks to the additional capabilities provided by the latest technologies, reinforces the military- and security-oriented approach to border and migration management. The most critical aspect of this arrangement is the ability of surveillance capitalism's business model to make projections and manipulate consumers, thanks to big-data analysis, used by states in border and migration management. This, in turn, makes it possible to identify migrants and allocate them based on workforce needs.

The role played by the overlap between these advanced techniques and the goals of intelligence agencies in making the business model of surveillance capitalism into an intrinsic component of contemporary capitalism is not something that happens behind closed doors. After 9/11, many countries, including the US, made it possible to access and monitor personal data via mass surveillance in the name of strengthening anti-terrorism legislation. We can talk about surveillance capitalism as a new, additional phenomenon that aids intelligence agencies in controlling societies, now partially outsourced to private companies. I would like to emphasize that this is not limited to tracking and monitoring only: forecasting and manipulation capabilities have also significantly expanded (Véliz, 2020).

What is important for our purposes is that technology monopolies that monetize surveillance, develop advanced technological solutions in this field and work closely with intelligence agencies also offer digitalization solutions to UN agencies and border and migration management departments of states. This background both shows the oligarchic grouping that forms and gives rise to concerns regarding the interest shown by this group of actors in migration and border management. As Véliz (2020) argues, 'Data is a toxic asset. Dangerous, sensitive, hard to keep safe, desired by many like criminals, intelligence, insurance companies'.

I prefer to use the term oligarchy to describe the business relations and alliances that manage the process of migrant selection through policy making, product development and use of surveillance applications. The main reason is that the process is shaped by a relatively small but very influential group of companies and bureaucratic agencies. Members of this group are in direct contact with one another thanks to business relations, lobbying activities and advisory boards, sectoral associations, fairs and so on. The process is conducted in secrecy because fields such as security, militarization and intelligence are considered sensitive. Hence, this is not a transparent or accountable process. Even though the rights of migrants and refugees are governed by international law within the framework of human rights, migration flows are the exclusive purview of security bureaucracy and are defined as a security issue, which makes it impossible to have transparency and accountability in terms of human rights. This can also be observed in the discussions and legal disputes between organizations defending migrants' rights and Frontex – the border security organization of the EU – regarding the prevention of irregular migration and pushbacks, which will be addressed in Chapter 3.

The states and global corporations in this oligarchical structure have positioned themselves as donors, using their economic and political clout, which renders UN agencies and humanitarian bodies accomplices in justifying the limits set by capitalism when it comes to individual migration movements. Therefore, if migrants from Syria, Afghanistan or Pakistan are designated to be employed in the supply chains of the textile industry in Turkey, for example, these organizations undertake work to this end, 'raising awareness' against migrants' desire to go to Europe. If, however, migrants insist on going to Europe despite these programmes, they are encountered by Frontex and national armies.

The mutual interests of the members of this oligarchic structure are discussed, with plenty of examples, in the following chapters. Military companies play their part in managing migration flows or cross-border labour movements, which is explored in Chapter 3. Many actors, from smartphone operators to financial firms, offer products and services to ensure that migrants join the labour force and turn into workers as soon as possible, which is analysed in the context of digital identities in Chapter 4. What is common to all these processes is that data analysis is conducted on migration movements, and ongoing data analysis deepens surveillance.

A critique of Zuboff's surveillance capitalism

I would like to use this section to share some of my criticisms regarding the theory of surveillance capitalism. Zuboff's (2019) surveillance capitalism approach is important in demonstrating how technology companies became monopolies and gained the power to predict and manipulate the future and undertake effective marketing. It is possible to take this as a starting point to interpret the cooperation that technology companies have with the military/intelligence industry and financial firms, as well as their investments in the field of migration and border management. However, this does not mean that I agree with all the points Zuboff makes about the business model of surveillance capitalism. This book was not written to evaluate and criticize the theory, but a brief discussion is in order.

For example, behavioural surplus is a controversial concept. I am not convinced by the argument that social media users generate surplus data and the analysis of this surplus data by companies amounts to the exploitation of users. I argue that the key concept here is rentier capitalism[2] (Christophers, 2022). These companies offer digital platforms. Users come together on these platforms usually to obtain services free of charge. The company that owns the digital platform in question brings these users together with other companies that would like to market their products and services. The advantage that these platforms offer, compared with conventional marketing companies, is that they can analyse big data using AI algorithms to predict the future, and show personalized/tailored ads to users that are predicted to show an interest in the service or product in question. In short, users are not creating value by undertaking uncompensated work; rather, they accept using a service free of charge in return for sharing their private lives and socializing on the platform, and the ads they are shown are selected based on their own posts and interactions. For the platform to be successful, companies using it for marketing purposes must be satisfied with the sales they can generate and other services they receive. Greater satisfaction translates into repeated and heavier advertising by the companies in question, and the platform shares in the surplus value of the advertisers. One thing to note about the power of marketing is that price and value do not have to equal one another in the market. In other words, the price of a product does not always match its value. The power of marketing comes from its ability to sell a product at a price exceeding its value. This turns digital platforms into a factor that limits the tendency of prices to go down, and digital marketing becomes a mainstream marketing method.

The defining characteristic of these digital platforms is their share in the surplus value obtained by companies doing business on the platform. So, a portion of all the profits generated by companies in all other sectors is transferred to these platforms thanks to their marketing power. Apart from protecting and strengthening the infrastructure for their platforms and improving the process of analysing personalized ads, Google or Meta do not really have to make special or continued efforts to innovate. All the innovation that matters already took place at the stage of creating the platform. Afterwards, they only needed to keep the platform liked and used by people voluntarily and the data flowing. This is why blitzscaling is important, which refers to increasing the number of users as quickly as possible, without expecting any profits in the first period and having investors/shareholders bear the loss. As user numbers grow and user engagement increases, more advertisers are attracted to the platform, paying a portion of their earnings as monopoly rent.

Reducing the issue to the exploitation of the behavioural surplus or posts generated by people would mean limiting our focus to the relationship between the user and the tech company. This, in turn, would prevent analysis of how tech firms intersected with all sectors and fields in the economy, gained a monopolist position and started to cooperate and integrate with official agencies.

However, apart from marketing and given the rentier nature of the platforms, they would attract interest from states, particularly from intelligence and security bureaucracies, for the advantages they offer in terms of user analysis and future predictions. They too would utilize the services offered by these platforms and make payments in return to control society, monitor enemies, predict the future and work more efficiently. The infrastructure offered by these platforms makes it possible to provide services to almost any sector, including finance and military industries. Cloud services and operating systems have now become infrastructure services needed by all organizations, whether public agencies or private companies. This is an important observation because, in addition to strengthening privacy as a way to fight companies that adopt surveillance capitalism as their business model, an alternative approach could be defended for the public ownership of these infrastructure systems and platforms.

- Thus, it becomes possible for tech companies such as Palantir and Microsoft – in the fields of migration and border management and humanitarian aid – to cooperate with military/intelligence/border-security bureaucracies and companies and provide technical support and platforms to projects that aim to limit migration movements;
- to provide infrastructure and platforms to UN agencies so that they can organize the distribution of aid to migrants; and

- to cooperate with financial firms to open bank accounts for hundreds of millions of people who have previously been outside the financial system.

Another aspect of digital/social media platforms and various applications is that they give capitalism the power to control leisure time, which reproduces the labour force. The most basic struggle under capitalism was the one undertaken to reduce working hours to limit the exploitation of the working class. To this end, workers in many countries took collective action, particularly via labour unions, and succeeded in reducing the legal working day to eight hours. Efforts to expand or shrink the working day are still an important part of class struggle today. Fewer working hours would mean less exploitation and more leisure time, enabling people to spend more time socializing and being with their families. More leisure time also means that working classes would have more opportunities to have a bigger say in politics. It is practically impossible for people who spend 12 to 14 hours daily in their workplaces to participate in politics, organize and make demands. The reduction of the working day to eight hours, on the other hand, created new opportunities for taking political and social action.

With social media and digital platforms, however, as millions of employees spend their non-working time in virtual environments as users, non-stop analysis of the data they generate becomes possible, as well as non-stop display of ads and messages. People may prefer to spend their time on their phones or computers, socializing among the millions of other people online – even though they may be alone and physically isolated – instead of spending time with family and friends, resting, reading, having hobbies, or taking social and political action. This, in turn, offers great opportunities to companies and states for social control and manipulation. Technology companies also offer digital environments for people who would like to take social and political action, organize, spend time with their families or read books, and convince everyone – who would otherwise be offline and outside the direct control of capitalism – to operate in online environments, using the infrastructure that they provide. However, sharing opinions or revealing details of one's personal life cannot be described as a process of production. In short, rather than describing the situation as the exploitation of behavioural surplus, I view it as an effort to keep users engaged with social media as long as possible, displaying them the posts that they would like for as long as possible (creating filter bubbles) and, as a result, having the opportunity to show more ads and political messages and mediate more sales, thus attracting more of the profits earned in other sectors.

Therefore, I do not mention 'the behavioural surplus' when discussing surveillance capitalism in migration and border management. Migrant people may use social media to arrange their journeys, receive the necessary information to integrate into the host society and could benefit from social networking in their leisure time. All the data collected and analysed from their social media accounts and CDR data from their mobile phones could be used by border agents in deciding their asylum or residence permit applications. For instance, in many EU countries, border agents confiscate asylum seekers' mobile phones, and happy photos from past celebrations might be used against them as they 'do not seem to flee persecution'. All the other surveillance technologies, such as sensors, drones and satellites, could be used by border agents to detect and track the mobility patterns of people on the move (Maxmen, 2019; Bircan, 2022). Here, states benefit from surveillance capitalism's power to surveil, analyse and predict. There is no critical concern, however, about behavioural surplus for migrants and refugees; this is not a personal relationship between a digital platform company and a migrant, and the company doesn't have the aim to exploit migrants' behavioural surplus. Surveillance technologies have a use-value for border agents or humanitarian organizations, and tech companies earn their revenues from serving these institutions.

On privacy

The technologies currently being tested on migrants and refugees (including lie detectors and facial recognition) recall to mind Zygmunt Bauman's (1998) quip, 'Great crimes often start from great ideas'. These technologies of course represent miraculous ideas and great technical progress. However, they are often used to serve an oligarchic structure made up of a small number of monopolist technology, finance and military companies and bureaucracies or political decision-makers in order to increase their profits and reinforce their monopoly positions. These technologies could also be used to improve people's living conditions, help solve global issues such as the climate crisis, decrease working hours or reduce inequalities, but this does not happen because of a small minority that claims private ownership over them.

On the one hand, the conditions are created by global powers that force people to turn into refugees or asylum seekers in pursuit of a better life, then, on the other hand, technologies are deployed to decide who will cross the borders and who will not. Examples include lie detectors, facial-recognition systems that target social movements and the right to protest, and business practices based on manipulation. Thanks to the latest technologies, this entire process happens in places distant from the states and actors, such as the Mexican desert, Mediterranean waters, Iranian mountains, Greek islands or a remote jail in Nauru, and, in Bauman's (1998) words, 'in a hygienic and remote manner, without unleashing the military or the police on people who do not abide by the "borders and rules" they create, nor letting blood or tears appear on our screens'.

We deal with issues of privacy and content multiple times in the following chapters dedicated to examining technological applications in big data, smart borders and digital identity. These are not simple repetitions, as they are meant to underline the shared aspects of technologies that are used for different purposes in migration and border management and humanitarian emergencies, as well as their relationship with surveillance capitalism. Even though some of these technological and innovative solutions force migrants to stop and stay in designated regions while others aim to contribute to migrants' social and financial integration, what they have in common is that they are all based on the rejection of the privacy of people on the move, be they legal or irregular (Latonero, 2019; Madianou, 2019). Achieving consent from people on the move or respecting their privacy is not even a consideration for states and monopolist companies that invest in these fields. This makes it easier to understand why people on the move are used as test subjects in the development of surveillance capitalism.

The defining characteristic of surveillance capitalism is to collect as much data on people as possible, and privacy can be an obstacle to achieving that (Zuboff, 2019; Véliz, 2020). In other words, if people are to resist surrendering their daily lives to surveillance capitalism, their first line of defence is to protect their privacy and demand that governments enact laws to reinforce privacy (although, we have seen how that demand would not be sufficient). It is, of course, possible for surveillance capitalism to overcome privacy regulations and develop new methods. One example is the continued violations of privacy in the EU despite the presence of strong laws such as GDPR. Another aspect is that privacy-oriented and cryptographic solutions (discussed in Chapter 4) are also developed by companies of surveillance capitalism, which ensures the continuity of data flows.

Because privacy is usually perceived as a personal issue, one of the most basic arguments used to overcome people's objections is that tons of data collected from billions of people would not have a personalized effect on individuals. Another claim is that people who just go about their lives and not, for example, commit crimes should not be bothered by having their data surveilled, given they have 'nothing to hide'. When it comes to people on the move, on the other hand, this issue is put in the context of the sovereign rights of states. People on the move, be they tourists, business travellers or asylum seekers, must meet certain demands by the state when they enter the territory of a sovereign country. Because fear and dislike of foreigners are widespread in societies (Bauman, 1998), the opposite is almost impossible. What is more, in cases of emergency, thinking about the privacy of people in need of humanitarian aid is perceived as a luxury we cannot afford.

In this sense, is privacy a 'first-world problem'? Is the demand for having limitations on the sharing of migrants' and refugees' data and respecting their privacy an unfair one? In her 2020 book *Privacy is Power*, Carissa Véliz rejects these approaches and argues that, as large corporations value the data that we produce, and compete with one another to obtain and process it, we should see how this is connected to power and capitalism. Weaker privacy means companies would be better able to track, predict and influence our areas of interest. The surveillance carried out by these companies is disempowering in the final analysis because it is not meant to protect or empower people or help them access services. In *The Age of Surveillance Capitalism*, Zuboff (2019) notes that tech companies that adopt the business model of surveillance capitalism do not like regulation or control and grow rapidly in deregulated environments. This shows that the rejection of privacy is a basic principle of surveillance capitalism.

Since privacy hurts the bottom line of surveillance capitalism, and privacy advocacy is one of the most important means of resistance against it, approaching the issue solely from the perspective of individual rights would not be sufficient. When we connect with governments, centres of power and the capitalist system, and understand the dangers of data extraction, it becomes clear that we cannot ignore migrants, refugees or people who work in mobility-based sectors. From this perspective, we need to undertake more effective interventions on behalf of vulnerable people and groups considered to be high risk in terms of humanitarian aid, because protecting them against the powerful is an important public duty.

However, when it comes to technological 'solutions' regarding migrants and refugees in the fields of data analysis, smart borders and digital identity, and the most speculative pilot projects run in many countries, it is clear that in places where regulation and supervision are much weaker compared with the countries of the Global North such as the USA and those in the EU, where principles of privacy do not apply to people categorized as foreigners, including migrants and refugees, and where privacy is considered meaningless or a luxury, conditions are much more suitable for developing and testing new products and adopting a more aggressive and carefree attitude.

Carissa Véliz (2020) summarizes this: 'World without privacy is dangerous. … Privacy is intimacy. We need privacy to think and make up our minds, to be free from unwanted pressure, and to be autonomous.' I would argue that, in making this point, Véliz was not thinking about people forced to leave their countries because they faced death threats or other risks; instead, she was addressing citizens of developed countries. Therefore, I find it important to call attention to the plans for and investment in the fields of migration, border management and humanitarian aid by actors of surveillance capitalism, that is to say, members of the oligarchic structure discussed throughout this book. It is not only important for protecting the rights of the people on the move, but it is also potentially very dangerous because of the risk that the technologies tested on migrant people could be unleashed on entire societies after a while.

Zuboff (2019) argues that deregulation is one of the most important steps in the development of the business model of surveillance capitalism. This is also in line with Bauman's (1998) observation that everything is deregulated and privatized in the postmodern world. Bauman describes this as: 'The universal deregulation – the unquestionable and unqualified priority awarded to the irrationality and moral blindness of market competition, the unbound freedom granted to capital and finance at the expense of all other freedoms'.

Tech companies implementing surveillance capitalism have turned, within a couple of years, into global companies and monopolies, making use of the opportunities provided by the neoliberal and postmodern environment. These companies can test their speculative and pilot products on migrants and refugees because privacy-oriented laws and other regulations – in developing countries and in the EU and US – are either non-existent or have very little teeth in the process of checking and selecting foreigners on the border.

The claimed – and sometimes real – ability to predict the future played an important role in the rapid development of the business model of surveillance capitalism, making it a critical component of capitalism today. Under normal conditions, the future is predicted to unfold in line with conclusions drawn from existing data sets. Conditions become even riper given the promise to deal with uncertainty – which is the biggest worry of companies and societies in the postmodern and neoliberal era and is reinforced with deregulation and privatization – and to resolve the fears and anxieties of a life that has become 'undecidable, uncontrollable and hence frightening' (Bauman, 1998). The digital platforms, AI algorithms and marketing techniques offered by these companies are in great demand because they not only utilize deregulation, which gives rise to these anxieties in the first place, they also promise people and other companies the ability to predict the future. In the face of the uncertainties created by the postmodern and liberal world, the faith that AI algorithms – which use advanced statistical tools to calculate probabilities – are telling the truth helps to deal with this fear.

Questioning the capitalist hegemony

The following chapters move on to an analysis of surveillance technologies used in the fields of migration and border management and humanitarian aid. They discuss the new technologies in the fields of big-data analysis, smart borders and digital identities; the reasons why these technologies are deployed; and the actors who develop these technologies. More importantly for this book's purposes, I will reveal the connections between these three fields. I have observed in existing literature that the connection between smart borders and digital identities is not questioned: a critical perspective is adopted regarding smart borders, while projects aiming to digitize identities are viewed more positively, or, at the very least, there is a hope that technical and political measures to be taken would be sufficient to solve any problems that might arise.

I have argued, both in the Introduction and in this chapter, that the use of surveillance technologies is not limited to the field of migration and border management, and it offers important clues for understanding the structure of contemporary capitalism and its future evolution. I believe that migration is key to understanding contemporary capitalism, and the concept of surveillance capitalism is crucial to understanding its future trajectory, and the goals of those who develop and implement these policies are not limited to migration and border management.

These efforts are not solely aimed at providing a theoretical framework. They have a practical importance as well to the subsequent chapters, providing cases regarding the questions of how to approach innovation in the fields of migration and border management and humanitarian aid, how to establish connections between innovative approaches in different sub-fields, and how to develop an alternative innovative approach.

My main emphasis here is that these efforts, involving some of the most powerful actors of capitalism – from companies that are monopolies in different sectors to advanced capitalist countries, and from UN agencies to humanitarian aid bodies and academia – have established a new capitalist hegemony in the field of migration. It would not do to disregard this capitalist hegemony and instead target the director of Frontex, the Greek government, Johnson, Trump, Orban or Erdoğan. Simply condemning a few weapons manufacturers or notorious tech companies like Palantir does not produce tangible results either. We need to observe the overall trajectory of capitalism and reveal how it is connected to surveillance in other social fields, warmongering and the dissolution of democratic institutions. If we are faced with a large and coordinated front, we need a similarly large front to respond to the current hegemony. It would not be limited to people on the move, because similar issues exist in many fields, from climate change to poverty, inequality and conflict.

The construction of a new capitalist hegemony in migration and border management takes place faster and more comprehensively thanks to surveillance technologies, without any democratic oversight, transparency or accountability. Because the new hegemony requires a new legal order, the legal framework developed after the Second World War is being denied, ignored or not implemented. As a result, statements like 'asylum should have been claimed in another country' become increasingly meaningless.

When we centre the discussion on capitalism, it becomes easier to make sense of the statistics that UN agencies and business associations love to post on social media. Statements such as, '1 billion people on Earth do not have IDs', '80 million people have been victims of forced migration', '100 million people are threatened by hunger' and '1 billion will have to migrate in a decade because of the climate crisis' underline the importance to capitalism of the fields of migration and border management and humanitarian aid. Societies may approach these numbers with panic or anxiety, but from the perspective of capitalism, this is viewed as an opportunity. Technological advances make it possible to steer the hundreds of millions of people whose lives have been upended and who have been dislocated, are on the move or potentially can move, thus shaping and positioning the working class at the global level. It functions, on the one hand, to push people who have been outside or on the margins of global capitalism's radar as workers or consumers to the centre, and, on the other hand, to reduce the demand for higher wages on the part of people who are currently caught in the machinery of capitalism as producers and consumers, denying their rights further and forcing them to take part in a fierce competition. The bonus is that people direct their anger at the people on the move, the newly arrived or the foreigners instead of at capitalism. Therefore, when we examine the examples in the following chapters in the context of a new capitalist hegemony, it will be possible to understand the process by which surveillance capitalism has transformed from a business model into a structural feature, and the growing interest of capital in this field.

Another practical outcome of this book is that we will have a better idea of how to evaluate innovative projects implemented in different fields and countries. For example, rejecting projects at the outset may be advisable if they involve companies that adopt the surveillance capitalist business model. Another recommendation might be to take care with projects that feed the capitalist hegemony discussed here, even if they would solve some of the main problems of migrants. This, in turn, may help develop innovative projects centred around migrants and workers and further their interests at all stages, from the design process to product development and implementation. The book also gives many examples of projects in the fields of big-data analysis and digital identity that were initiated with the best interests of migrants in mind. However, as we will see, some of these generated results incompatible with the initial aim of migrants' integration and empowerment. Therefore, one of my aims in this study is to contribute to the emergence of alternative, counter-hegemonic approaches.

2

Migration and (Big) Data Analysis

The analysis of (big) data has exciting potential in the fields of migration and border management and humanitarian emergencies, as in many other fields. The reason I place 'big' in parentheses is that the definition of big data keeps expanding in quantitative terms. Therefore, it might be sufficient to simply talk about data analysis. Two issues make data analysis especially valuable. First, data sets that are too large for humans to analyse unassisted can be evaluated using AI algorithms thanks to advanced statistical applications. Second, with advances in cloud storage, it becomes possible to hold and use large data sets, without data going to waste.

What makes this analysis exciting, specifically for the issue at hand, is that inferences made from data sets allow us to analyse the basic characteristics and dynamics of mass movements and make predictions about the future (Bozçağa and Cansuna, 2022; Gürkan et al, 2022; Solano, 2022; Weber, 2022). The business model of surveillance capitalism, discussed in Chapter 1, is also based on the assumption that similar phenomena, under normal conditions, should lead to similar expectations. This, in turn, offers new opportunities for making projections about migration movements and analysing applicants to cross borders. Examining data from previous applicants and categorizing new applications using this framework make it possible to make judgements about whom to allow to migrate and where.

Capitalist companies' dominance over big data

The biggest concern, for the purposes of this study, is that a significant portion of the main actors that collect and analyse data from migrants are capitalist companies. States, UN agencies and humanitarian aid organizations also collect data from migrants and refugees, but the field is dominated by a small number of companies. Mobile phone operators, financial firms, digital platforms, social media providers and companies that operate search engines and offer translation and navigation apps collect real-time data from migrants on a continuous basis. Using this data, they offer products and services to practically everyone, regardless of whether they are migrants are not. For example, as a Turkish-speaking migrant in the UK, I am usually shown ads for remittance companies when I use social media or search engines. Moreover, because many of these companies operate at the global level, they have the means to monitor, in real time, developments taking place in multiple countries experiencing migration and refugee movements. Companies such as Vodafone, Orange, Mastercard, HSBC, Facebook/Meta and Google have the means to analyse migrants in multiple countries, and even to track certain migration movements (Weber, 2022).

I have conducted interviews with officials from mobile operators in Turkey, and some of them, speaking on the condition of anonymity, said that they shared data on refugees with security and intelligence units on a daily basis. Similarly, there are claims that refugee data was shared with the government of Bangladesh, or with one of the warring sides in Yemen. These are examples of the generally accepted fear that companies share their data with the security and intelligence units of states.

What I would like to emphasize in this section is that private companies' dominance over big data analysis should be questioned more vigorously. The problem here is that companies that collect and analyse data have the capacity to monitor and direct 'people on the move' better than states and UN agencies. Therefore, many researchers have the hope and the expectation that these companies would share some of these data sets with researchers for academic purposes without questioning their dominance (Bircan and Korkmaz, 2021).

It is not possible for individuals to accurately guess what sources are used to collect data about them. Data collection is not limited to social media and other digital platforms and can include many instruments, from the internet of things to satellites (Bircan, 2022). People on the move are not able to recognize the data points about themselves being collected and so take appropriate counter-measures. This issue extends to questions regarding what happens to the data once it is collected, who processes it and with whom it is shared. These are questions that tend to remain unanswered. One of the more secretive players in this field are data brokers, intermediary organizations that buy and sell data. Because they are not often publicly visible, I believe companies that operate as data brokers in regions with high levels of migration and international mobility should be put under closer scrutiny.

This applies to everyone, regardless of whether they are migrants or not, but, because of the particular vulnerability of the migrant experience, manipulation, disinformation and fake news – all a normal part of daily life – can have life-or-death consequences for migrants. In particular, the principle of no harm – which must be abided by states, UN agencies and aid organizations in the areas of forced migration, humanitarian aid and refugees – becomes meaningless for profit-seeking private companies that do not have any responsibilities in this regard. Their priority is to turn people into labourers, customers, data sources, consumers and producers of surplus value. Demands or principles involving anything other than these are treated as meaningless and disregarded. This process is also consistent with the capitalist hegemony established in the field of migration and border management. The might of the pro-market approach, which is paramount in the field of data collection and analysis, provides clues to the process by which migrants are steered and the working class is reshaped and repositioned at the global level.

Compared with monopolistic companies, the security, military and intelligence bureaucracies of advanced capitalist states, which dominate the field of border and migration management, also abide by this process. They view migration as a security threat that needs to be managed, and intelligence as playing a critical role in overcoming security issues. Therefore, access to data is vital. In this regard, analysis of social media platforms and cooperation with mobile operators are important, but different data sets are also included in the analysis thanks to smart border products. A wide variety of technological products, such as satellite images, border sensors, robot dogs, drones and fake phone towers, are used to collect data on the borders for analysis. This, in turn, involves companies that produce specialized military products or those working in surveillance environments. Therefore, we are faced with a process that is based on a pro-market approach and monopolistic companies collecting data from migrants and refugees, but which is coordinated in the background by the security-oriented and militaristic approach of the state.

Parameters set by the security-oriented approach also apply when monitoring (via biometrics and digital identities, for example) those allowed to cross the border and their process of adaptation to the host society. For example, financial firms and mobile operators are also included in the surveillance system via 'know your customer' requirements that apply when purchasing a SIM card or opening a bank account, or via requirements concerning money laundering and terrorist financing in the case of financial transactions (Slavin, 2021), which sheds light on big data's relationship with the security-oriented approach to migration management and with the oligarchic structure that forms. We will discuss this in Chapter 4.

The data question in migration

Data sets used for data analysis and machine learning in migration and border management come in many forms and shapes, but they come from two main sources. The first is making use of historical data sets. This includes different sets of data previously collected by official agencies or shared by people applying for visas or residence permits. These data sets are used to categorize people on the move and calculate their risk scores to prevent or allow them to cross borders. The other source is the analysis of real-time data. Satellite and drone images, sensors and other products of the internet of things, social media and smartphones generate lots of data about people on the move, whether they are aware of it or not. Thus, both historical data, that is to say, data produced and collected in the past, and momentary data, collected in real time, are used to position security forces on the border, decide whom to allow across it, and keep up the surveillance of people on the move.

People on the move comprise a very large group. Even if we were to focus solely on forced migrants and refugees, there are tens of millions of people that are in direct communication with and receive aid from organizations such as UNHCR, WFP and IOM. Millions of other people also migrate legally every year, categorized as tourists, students or workers. In addition, there are hundreds of millions if not billions of people who live in countries affected by mass movements of asylum seekers and forced migrants and who share the same problems, such as poverty (for example, the Rohingya people seeking asylum in Bangladesh). Therefore, when we talk about migration and mobility, we are talking about very large groups of people who would be of interest for contemporary capitalism.

It is also notable that strong legal protections do not exist regarding privacy and consent in the countries that are on migration routes for a significant portion of the people on the move. There is a widespread perception that migration flows originate from the less-economically-developed regions of the world, the Global South, and end in capitalist centres or the Global North, but a significant portion of these flows actually take place among Global South countries. Countries that are most affected by flows of refugees and forced migrants and that host the largest numbers of refugees are also poor countries themselves. These countries tend to have weak and insufficient laws for the protection of the data generated by people on the move or even by their own citizens. Even when legal protections do exist, many countries are unable to enforce them against tech giants and issue fines or negotiate.

The capacity of many African and Asian states to collect data from people living in their territories is weaker compared with international corporations operating in these countries. While many states struggle with conducting a census of their population because they are unable to allocate sufficient resources or employ a sufficient number of people, phone operators are able to collect a range of real-time data, on a daily basis, about a significant proportion of the population as most people have smartphones. These companies can access and analyse all sorts of data, from people's spending habits to changes in their economic situation, and from their levels of education, commercial activities and employment status to their personal lives (GSMA, 2018; Bozçağa and Cansuna, 2022). Internet access may be low compared with developed countries, but social media companies, digital platforms and financial institutions can still access valuable data. Moreover, global corporations have greater opportunities for control and surveillance compared with a majority of less-developed countries as they offer mobile banking and payment services, bring different sectors together and control overseas payments such as remittance transfers. This, in turn, creates a lopsided relationship that reproduces and reinforces the dependent position of these countries in the capitalist system and ensures the continuity of colonial ties. In short, given that regions that are keen on protecting personal data and ensuring privacy – such as the EU – also face serious problems in auditing and sanctioning corporations, it would not be reasonable to expect emerging countries, which are home to billions of people, to be successful in this regard.

The effects of investments and policies regarding migrants and refugees – which are aimed at reaching new markets and expanding the influence of capitalism – are not limited to people on the move, and affect locals as well, because a significant proportion of human mobility takes place among the less-economically-developed countries. Developed capitalist states make great efforts to limit migration flows into their countries, and factors that trigger migration movements in many regions of the world (such as wars and conflicts, natural disasters and climate crisis) show no signs of abating. For example, when a smartphone operator makes an investment to provide financial aid to people forced to migrate because of a conflict or natural disaster in an African or Asian country, cooperates with the UN, receives funding, bureaucratic and academic support from advanced capitalist countries and makes an agreement with financial institutions, this benefits not only the newcomers but also the local communities who suffer from similar problems. The aid policy is designed to cover existing populations in the region as well, so as not to give rise to new problems among communities, and even when aid policies are specifically targeted at migrants, similar services, products and solutions spread into the local community, whose members witness these developments.

Because of the widespread availability of smartphones and active use of social media, and because mobile banking is actively used for both domestic and international payments, many companies, including technology, finance and telecoms firms, have become the biggest collectors and analysts of data in world regions where under-developed countries are concentrated. As most of these companies rely on the business model of surveillance capitalism, their goal is to serve customized ads and messages by analysing data, categorizing masses of people and making projections of the future. This, in turn, provides invaluable resources to capitalism for the surveillance, manipulation and steering of people in a given direction. Humanitarian associations collaborate with these companies to provide humanitarian support and in doing so expand these companies' spheres of influence and protect their hegemonies.

Surveillance of individuals and societies is not limited to the capacities of these companies. They are complemented with satellite technologies and other conventional tools of control and intervention used by military and intelligence bureaucracies and industries, making it possible to monitor vast regions and large movements of people. Moreover, as discussed in Chapter 3, some of the companies work with military bureaucracies to develop technological products called smart border applications, whereas some others contribute to migrants' financial and economic integration into their new countries, which is basically shorthand for their participation in the labour force. Therefore, it is quite possible to analyse, in real time and on a continuous basis, the movements of people from one end of the world to another, encourage or discourage potential migrants, and make calculations to decide how many of them to allow and where. This coordination is made possible by the size and reliability of the data sets in question, the technical capacity and technological products available, and the cooperation between monopolistic companies and security bureaucracies, compared to an oligarchy in Chapter 1. The main motivation behind this cooperation, as also discussed in Chapter 1, is to reshape and reposition the global working class, deepen the process of surplus value extraction, prevent opposition by the local working class, and turn large masses of people, who are currently on the margins of capitalism, into workers, producers, consumers and data sources.

These large and valuable data sets, which are not subject to any real oversight, are not limited to concrete products and services either. According to a GSMA report (2018), migrants make greater use of smartphones than locals, call their acquaintances in other cities and countries more often, use social media and visual and written means of communication more actively. So, as a result, offering products and services that cater to migrants is not simply a matter of social responsibility for mobile operators; it is also a lucrative business opportunity.

On the other hand, some might ask, what is the harm in turning people into customers or workers if they used to rely on social aid and faced hunger and starvation, or how much profit can you really make out of people who live in poverty? Of course, in these cases, the main clients of technology, finance or telecoms companies are not the people in need, called beneficiaries, they are the UN agencies and humanitarian aid bodies that have a responsibility to help migrant people. These organizations in turn use funds received from donor countries to make payments to the companies in return for their services. However, the data collected is quite valuable in and of itself, apart from such arrangements, because having access to data from large groups of people – who do not feature in official records and are considered to be untouched communities for global capitalism – plays an important role in training algorithms. Therefore, a narrow view of profit and loss would not be sufficient to make a proper assessment of the situation.

The humanitarian aid industry as a data supplier to technology companies

We have reached an interesting point in our discussion. Because data has become valuable in and of itself, and because this field involves tens of millions or even hundreds of millions of people, the nature of the relationship between UN agencies and humanitarian aid organizations on the one hand and companies that adopt the business model of surveillance capitalism on the other has changed. As digitization makes it possible to provide services to large numbers of people quickly, utilization of data science, AI, mobile banking and blockchain is encouraged by donor countries in order to have a more transparent and accountable process and carry out fast and effective interventions (Coppi and Fast, 2019). However, UN agencies and humanitarian aid bodies do not have the technical capacity to deal with big data of these proportions, therefore cannot act like financial companies with the ability to facilitate mobile payments throughout the world, and do not have the technical capacity or human resources to develop these technologies from scratch for themselves. Crucially, such organizations do not have the budget to do so. Donor countries encourage these organizations to make use of the latest technologies, but the budgets they allocate or timescales they expect do not allow for development of large-scale technologies or employing people that would be required for such projects.

What is more, many donors that request innovative solutions to be developed require working with tech companies from their own countries to do so – indeed, this is often necessitated by capacity demands and budget limitations without private-sector collaboration. For all these reasons, aid organizations find themselves working with tech companies. The cooperation is not limited to offering solutions for specific products and services. For example, offering cash transfers via mobile banking and providing aid to 80 million to 100 million beneficiaries around the world, on a regular basis, and doing so in a transparent and accountable manner requires having a robust data-management system.

The point made in Chapter 1 applies to big data analysis as well: public authorities used to be the most influential actors in issues around migration and refugee movements because profit-seeking was not seen as right, legitimate or ethical when it comes to refugees, but the world has changed and this approach is being replaced by another in which corporations are the most influential actors, and the field is dominated by corporate agenda-setting and the political concerns of security bureaucracy. When public authorities were the most influential actors, the role of the private sector was limited to supplying the required products and services. However, as a result of the pro-market and corporate-oriented approach, aid organizations are turning into data suppliers for technology, finance and telecoms companies (Hosein and Nyst, 2013; Taylor and Meissner, 2020).

As will be discussed with examples in Chapter 4, representatives of UN agencies and aid organizations on the ground are expected, when they arrive in war zones, poor regions or areas stricken by natural disasters, to record aid recipients and verify their identities, making sure that they really are who they claim to be. Moreover, they are asked to collect data using the questionnaires they are given. They are also asked to obtain consent from beneficiaries for the regular collection of their data and explain and promote the technological solutions offered. Indeed, for the beneficiaries to be able to receive regular aid, or to receive cash transfers via mobile banking, for example, SIM cards or prepaid cards have to be distributed first. Therefore, at the end of this busy process and despite the often hectic and life-and-death situations encountered in humanitarian emergencies – in an environment where technical and communication opportunities are limited and where panic, uncertainty and fear occasionally rule – aid workers complete the registration and verification process, distribute SIM cards and phones, and help open accounts for mobile banking so that the process is completed and these people appear on the radar of capitalist companies and start generating data using their phones.

This entire process takes place under the most unsuitable and problematic conditions. We encounter these practices being implemented in conflict situations in Yemen, Syria and Afghanistan, among Rohingya refugees in Bangladesh, among Syrian refugees in Jordan, Lebanon and Turkey, among people who flee conflicts and natural disasters in Nigeria and Kenya, and among earthquake victims in Nepal. On the one hand, it is obvious that opportunities such as having SIM cards and bank accounts are important for receiving financial aid, as well as for communicating and participating in social and economic life. In a world where wars and conflicts last forever and it takes many years for countries to recover after a natural disaster, having rapid access to these tools certainly provides great opportunities. On the other hand, however, the humanitarian organizations that undertake these efforts under the most difficult conditions end up supplying data to private companies and expanding their network of customers. There is no reason to ignore this aspect of their work. Many companies are enjoying great opportunities for their business model of surveillance capitalism without having to dispatch their own personnel to war-torn Yemen and Afghanistan, or to Haiti and Nepal after the earthquakes there, using public funds and thanks to the work of aid organizations operating in these regions; and all this in a world where having access to data sets is invaluable. This, in turn, damages the originality, autonomy and public nature of humanitarian organizations, which are under pressure from corporate agenda-setting and lobbying and given directives by donor countries, and have to operate within the narrow framework set by security bureaucracies that view the entire process in terms of security risks and threats.

In this context, controversies surrounding the deal that the World Food Programme made with Palantir and the disagreement that the UN agency had with the Houthi government in Yemen are instructive (Greenwood, 2019; World Food Programme, 2019a, 2019b; Parker, 2019; Responsibledata, 2019). The WFP is one of the biggest UN agencies, and provides regular food and cash aid to close to 100 million people around the world, a figure that keeps growing. The WFP deals with issues such as the spread of wars and conflicts, climate crisis and the resulting food insecurities. Having regular access to 100 million people around the world and making efficient and transparent use of the funds provided by donors require a robust data-management system and technical infrastructure. To this end, the WFP made a deal in 2019 with Palantir, one of the leading data-mining companies in Silicon Valley. Palantir produces surveillance technologies and offers products and services for border security, intelligence and military uses, works with the CIA and the Pentagon, has a boss that served as an adviser to President Trump, and is notorious in the eyes of critical academics and NGOs. Thus, a company that had a well-known political engagement, developed technologies that led many migrant people to their deaths, and worked closely with the US military and intelligence establishment gained access to data collected from close to 100 million beneficiaries of a UN agency, which was supposed to be impartial, and this drew a strong negative reaction from many organizations.

One aspect of the issue is that gaining access to data of millions of people from world regions that have strategic value and where different imperialist powers jostle for influence is a highly sensitive subject, and it is only natural for a controversy to arise when this access is provided to a company known for its anti-immigrant and far-right views and links with the CIA. Even if the company promises not to use this data for any other purposes, it would be difficult to take its word for it. Moreover, the aid agencies involved typically do not have permanent audit teams.

However, a UN agency cannot be expected to compile a blacklist of private companies. It is also true that there is only a limited number of companies that can offer a data-management system to the required specifications, so the company that made the best offer would likely be selected at the end of the tender process.

On a side note, I could make a speculative argument here. I could argue that a Chinese technology company such as Huawei would not be granted the contract even if it were to offer a similar system under much better conditions (Benner, 2020). Yet, it would not be possible to exclude Palantir from the tender in an environment governed by rules. On the other hand, would it really be better if Microsoft or IBM were granted the contract, and what would be different if this were the decision made? Other leading tech companies like these also work with the Pentagon and the wider security establishment in the US. For example, companies including IBM, Microsoft and Amazon competed for a contract to offer cloud services to the Pentagon, and the contract was awarded to Amazon (Delcan, 2019). These companies are discussed in more detail in Chapter 3. I would argue that they would not draw as much criticism as Palantir did. This, in turn, points to another problem. Gaining access to data from millions of people who are either outside or on the margins of capitalism is a great opportunity for monopolistic companies that operate in the field of data analysis, adopt the business model of surveillance capitalism and engage in data mining.

An alternative would be for UN agencies to have the in-house capacity to develop their own data-related technologies, adopting an independent and long-term approach. That the donor countries do not provide them with the funds required to do so is consistent with the overall argument made in this book.

Another controversy about the WFP stemmed from the dispute it had in 2019 with the Houthi government in Yemen. The government asked people to refrain from providing their data to the WFP. It claimed that data collected by the UN agency as it distributed aid was obtained by Saudi Arabia – which provided active support to the opposition in the Yemen civil war – and that Saudi planes selected their bombing targets partially on the basis of this data. The first reaction of the WFP was to pause the distribution of humanitarian aid. The two sides came to an agreement after negotiations, and aid distribution was resumed. However, independently of the controversy in question, these claims and suspicions point to the risks of large corporations in the field of data management and analysis getting involved in the process when they have clear connections with governments and security agencies. It also means that people who work on the ground and distribute aid under difficult conditions end up playing a role in the collection and transfer of data, regardless of their intentions.

The humanitarian aid sector transformed in the grip of capitalist hegemony

If we are talking about forced migration and refugees, and wars or conflicts are involved, and if we are claiming that people on the move are in a vulnerable position, it is only natural for intelligence agencies to be interested in this work. Analysis of the data collected would naturally provide advantages to the warring sides and their supporters. This is because issues such as the controversies surrounding the WFP, the UNHCR providing data on Rohingya refugees to the government of Bangladesh, and data collected by the US on people in Afghanistan ending up in the hands of the Taliban shed light on the debates around the importance and limitations of data collection and storage. In this context, the hacking of the IFRC (International Federation of Red Cross and Red Crescent Societies) in January 2022 should serve as an important warning. The IFRC is one of the top organizations in the humanitarian aid sector when it comes to planning for and investing in data security, as it collects data from people who live in the most volatile regions of the world and are in a most vulnerable position. It was nevertheless the target of a sophisticated cyberattack and had data on 550,000 people stolen. Given the level of sophistication of this cyberattack, the IFRC claimed that a state actor was behind the attack, which shows that we have to analyse the interest of official security and intelligence agencies and monopolistic companies in the fields of humanitarian aid, forced migration and refugees.

The UNHCR, WFP and IFRC are among the top organizations in this sector in terms of investing in data management and cybersecurity and cooperating with one another. However, there are many more organizations operating in the field of humanitarian aid, and many do not have the experienced personnel or the resources to devote to data-management technologies. Yet, they have to undertake 'innovative' projects in order to attract attention and funds from donors. Because most projects in humanitarian aid are short term and there is a high turnover as many employees work on a short-term basis, people working on one project often do not know about the fate of the data collected in another project. For example, in 2021 I conducted an interview with the director of a medium-size technology company that provides technical infrastructure (from data management to blockchain) to more than 30 organizations operating in humanitarian aid, who said that the company kept data on close to 100 million beneficiaries from more than 40 countries on its servers. The company provided the technical infrastructure to projects in African and Asian countries. It stated that it collected and stored all data abiding by GDPR rules, kept the data for as long as the laws and the contracts required, very few people in the company had access to the data, and it did not conduct any extra analyses on the data. There is no way of knowing this to be true. Legal and regulatory frameworks, and relevant checks and balances, simply do not exist in this space. Moreover, it is up to the company in case of a potential breakdown or attack whether to make this public or not. In short, data collected on millions of people and from many projects completed over the years by local and international NGOs and humanitarian aid organizations of all sizes may be left at the mercy of companies because of a lack of technical infrastructure and experts and because the projects are short term.

A private sector-oriented approach is apparent in this case as well. If it were not, 'donors' could have already helped organizations operating in humanitarian aid to improve their capacity in terms of technical infrastructure and expert personnel, and had the projects formulated in a way to encourage long-term employment. However, it is clear that this is not preferred, and this approach is consistent with the market-oriented new capitalist hegemony and the oligarchic approach of security bureaucracies and a few monopolistic companies.

The most important consequence of this is that aid organizations and other NGOs turn to the donors and compete with one another to get their approval and funds. The result is that they become organizations with employees who work on short-term contracts, make sacrifices, face a lot of stress, and are motivated by the spirit of humanitarian aid, coupled with CEOs in headquarters who are focused on satisfying donors' requests, do not have the skills to create other financial sources, and run the organizations as project-management firms rather than as NGOs. Migrants and refugees are then defined as 'beneficiaries', which does not leave room for an egalitarian approach based on solidarity, one that would give the initiative to migrants themselves and improve democratic representation. Refugees who live in camps for many years, and work for many years in camp workshops but cannot gain the status of legal workers with social insurance and the right to join unions, are seen in many counties from Jordan to Bangladesh and from Uganda to Kenya. This, in turn, is inconsistent with a participatory approach that would include migrants in the process as much as possible, allow them to realize their potential, and reinforce the notion of people making their own decisions. Instead, it ends up limiting responsibility to keeping the promises made to the donors, with a focus on quantitative measures of participation, being content with ticking off tasks on a list, and viewing transparency and accountability as being owed to donors rather than the migrants.

There must be a growing chasm between headquarters filled by professionals and run by CEOs on the one hand, and, on the other, local workers who brave the hardships and dangers on the ground, working under the most difficult conditions, motivated by altruism, but who end up being exploited by short-term contracts and disenchanted with the spirit of humanitarian activism (Duffield, 2012). Failure to allocate long-term resources hinders the development of cooperative and egalitarian alternative approaches.

Academic debates on big data and migration

Overall, the academic literature arguably has an optimistic perspective on data analysis. There are, however, access issues for academics studying sets of big data. Many academics are making great efforts to persuade collectors of this data to share it after anonymization and under suitable conditions. Campaigns to increase the number of open data sources and efforts to ensure that the sources comply with best practices regarding ethics and the protection of personal data also emanate from this optimistic perspective (R. Burns, 2015; Meier, 2015; Beduschi, 2017; Dave, 2017).

When a number of data sets on migration and mobility are available, many academic groups examine these from different perspectives. For example, predicting migration movements, analysing the social and economic integration of migrants, examining the connections between migrants and locals, identifying the regions that would facilitate migrants' integration into the host society and developing early-warning mechanisms to recognize the risk of conflict between migrants and locals before it is too late are some of the uses to which data analysis can be put in the fields of migration, border management and humanitarianism. I have outlined many concrete examples of (big) data analysis in my previous works, including *Data Science for Migration and Mobility* (2022), co-authored with Albert Ali Salah and Tuba Bircan, but I would like to call attention to three main issues here.

Concerns about data analysis in studying migration movements

First, a significant portion of the academic studies that use big-data analysis to examine migration and mobility display a positive and sympathetic attitude to migrants in their aims and motivations for their work. At the very least, the aims mentioned in the academic publications I have seen do not include preventing migration, or making people regret their decision to come and turning them into cautionary tales for others who might consider migrating. The aims do include, however, faster integration of migrants, better preparation by authorities so that they can improve the process by which migrants' basic needs such as education, healthcare and housing are met, and timely identification of potential problems. However, just as the road to hell is paved with good intentions, these analyses may generate the opposite outcomes. The biggest reason is that these studies target authorities (governments, bureaucracies, local governments and in some cases UN agencies) as their main audience. Academic research groups are not conducting these analyses so that they can be shared with the migrants themselves; instead, almost all are meant to be seen and examined by authorities, who would hopefully act based on this 'objective' knowledge. The naivety here stems from the assumption that at the root of migrants' problems lies authorities' inability to make the required political decisions because they are unable to access objective and detailed information.

As discussed in more detail in Chapter 3, however, applications that identify migrants and force many of them to stop before arriving at the border, and leave them to die if they persist with their journey, are also based on big-data analysis. Almost all the decision-makers who are the intended recipients of these analyses are already intent on stopping migrant and refugee flows.

Naturally, when they receive news of a potential migration movement, their first reaction would probably not be to make preparations to have sufficient food and shelter; instead, they would start building walls and mobilizing security forces. Although they do conduct data analysis, usually on legal application procedures to attract the desired workforce, they do not display a sympathetic attitude toward people described as refugees or irregular migrants, who attract more attention from academics. As a result, these analyses may increase the risk that more migrants would die.

At the same time, many academics who conduct data analysis in their studies on migration and mobility tend not to collaborate with migration scholars, which creates other problems. An academic may be fascinated by the data set they analyse and the statistical methods or tools they employ, failing to take into consideration the potential social and economic problems that their findings may lead to, and the possibility that the data in question may have sensitive and crucial meanings for the generators of these data. It is obvious that studies of this sort require interdisciplinary cooperation because any mistakes or deficiencies can have the potential to affect human lives (Crisp, 2018). Thus, some studies use phrases such as 'the assimilation of migrants' in a positive sense, although it is criticized in migration studies, and note that their aims do not include supporting assimilation. This might be simply a problem of word/concept choice.

However, theories such as 'new economics of labour migration' argue that the decision to migrate is taken as a family and not by individual migrants, as the family decides some of its members should migrate in order to diversify the economic resources available to the family or serve as a sort of social insurance. In migrations from Central Asia and the Caucasus to Russia and Turkey, from African countries to Europe, and from Latin America to the US, as well as in Filipino women's migration to work in domestic services, the desire to have at least one member of the family working overseas and sending regular remittances plays an important role. When these specifics are not known, analysis of data collected from migrants and comments on migration movements would be lacking.

In another example, data analysis that aims to have migrants employed as soon as possible (discussed in Chapter 1) focuses on matching migrants, using the data collected from them, with potential employment opportunities in the localities they may settle in. This, in turn, may result in it recommending jobs that do not require linguistic skills. Even when these studies are 'successful', the data analysis does not take into account the social and cultural lives of migrants, the trauma and stress they face and their expectations for the future.

Yet another example is the recommendation that migrants be settled in cities and towns that are sympathetic to their arrival, in order to facilitate social integration (Bansak et al, 2018; Calderon et al, 2022), but recommendations of this sort can be counterproductive. It might seem reasonable enough to think that conducting surveys to identify the cities that are the most sympathetic toward migrants and refugees, and having the migrants settle in these localities, would lead to establishing a quick dialogue with the local community, which in turn would facilitate the process of integration. However, what matters is the political content of sympathy or opposition toward migrants, and it is essential to recognize that such attitudes are not fixed; they are dynamic and amenable to change. Localities initially sympathetic to migrants may change their attitudes following large migration flows, and anti-migrant attitudes may become dominant. Or, countries that have been identified, on the basis of surveys and data sets, as having negative attitudes toward migrants may experience a change of heart, as might have happened in Poland following its growing sympathy for the Ukrainian people since the war there began.

In short, data analysis in migration and mobility does not consist simply of accessing data sets, conducting successful analyses using advanced statistical techniques and reporting the results. Migration movements can have diverse motivations and particularities. It is crucial to conduct analyses with migration experts and utilize the accumulated wisdom in the literature on migration.

Digital inequalities and ethical concerns

In the analysis of real-time data, in particular, digital inequality is becoming a significant obstacle. For example, many field studies call attention to the heavy use of smartphones by migrant people. New arrivals to the Greek islands in the Aegean are known to ask for SIM cards for their phones before asking for food and drink (Latonero et al, 2018). Mobile apps make it possible to navigate, use translation services, ask for help during emergencies, contact smugglers when needed, stay in touch with loved ones, and perform financial transactions. Active and frequent use of social media platforms, for the purposes outlined, in addition to learning about previous arrivals' experiences, provides migrant people with important opportunities in terms of understanding the journey and the conditions in the new country following the journey, as well as solving potential problems along the way.

This chapter is concerned with the opportunities that the digital/digitalized aspects of a migrant's journey creates for surveillance by states and corporations. However, it is important also to stress, at this point, that there are digital inequalities *within* migrant populations. Digital inequality refers to differences in terms of owning or having access to digital tools and opportunities. Many studies show that a significant portion of asylum seekers, irregular migrants and, in particular, low-skilled migrants have little or no digital access. Women are also known to have less access to digital tools compared with men. Class- and gender-based factors play an important role in creating digital inequalities among people on the move. Field studies on migration movements from Afghanistan and South and Central America report that a significant portion of the masses that migrate do not have phones, and cannot, for example, follow the latest news.

Important differences also exist in terms of the digital tools that migrants can use. Just as having or not having a smartphone is an indicator of inequality, as is having a higher- or lower-quality phone. A significant portion of migrants do not have smartphones and so are unable to utilize many crucial apps. Moreover, many migrants do not have access even to 3G. Data generated by migrants who can only use 2G, to make calls and send text messages, would be limited.

In some contexts, migrants make a conscious choice to avoid using mobile phones in certain regions to overcome surveillance, preferring to be invisible to the authorities. Some migrants are known to refuse to stay in refugee camps or apply to aid organizations so that they would not be forced to give up their biometric data or be registered.

Another aspect of digital inequality that needs to be considered is that data sets are used in the field of humanitarian aid and emergency, and aid is distributed on the basis of data collected via crowdsourcing (Meier, 2011, 2015; Jacobsen, 2015; Mercy Corps, 2016). For example, following the earthquakes in Haiti and Nepal, aid providers cooperated with mobile operators on issues such as where people waiting for aid were concentrated, and earthquake victims sent text messages to help with directing aid workers. It has been argued that this is a bottom-up approach that deals with grassroots organizations and makes it possible to communicate directly with aid recipients. However, it is also true that people who had smartphones, had sufficient credit on their plans, and were able to charge their phones following the natural disasters in their countries were, more often than not, members of social groups with higher socio-economic status. It is much more difficult for the poorest people – who may have been hardest hit by the disaster – to learn about these processes and make themselves heard by humanitarian bodies, thus impacting the data collected and causing any model or tool to be skewed in favour of the higher socio-economic status. Directing and concentrating aid on the basis of data coming from the field may result in the exclusion of many social groups. It is argued that this is similar to dealing with a disaster in New York with all the data coming only from Manhattan (Latonero et al, 2018).

I have mentioned only a few of the many obstacles to accessing digital tools as a migrant. Importantly, these digital inequalities are crucial in migration studies, and researchers should be aware of that when conducting analyses on data collected from migrants, as the data may not represent all migrants.

Abiding by ethical principles, in particular, making special efforts to protect those in a vulnerable position, is an important issue in migration studies.

This is reflected in principles in the fields of big data and AI regarding the anonymization of data, protection of personal data and consent. However, analysis of migrants based on data sets arguably does not involve as much care and diligence compared with the work of researchers who conduct one-to-one interviews with migrants. When you conduct in-person interviews with migrants, develop rapport with them, obtain their consent and then share the findings of your study, you address someone directly, and trust plays an important role for both the migrant and the researcher. However, in data analysis, the researcher does not have any real connection or communication with the migrants who generate the data. The main reason is that the data is not directly obtained from the migrants themselves, instead, it is provided by a corporation or an organization. These organizations, in turn, operate within the confines of the legal frameworks that bind them, and not necessarily with the best interests of the migrants in mind. They deliver the data set accordingly, after making interventions in line with their organizational interests. Researchers, therefore, do not have any apparent role or responsibility in the process at all, from the anonymization of data to obtaining consent. And the consent in question may only consist – if the data set came from a phone operator, for example – of a few lines stating that the customer accepts the sharing of their data with third parties, in a terms-and-conditions document of tens of pages agreed to when purchasing a SIM card. This casual, careless and irresponsible attitude disregards the issue of whether the data analysis to be conducted can bring harm to the people involved.

Furthermore, many studies on AI and data science show that absolute anonymity is not possible, and various techniques can be used to trace data back to individual people. Both the variety of data sources and the variety and number of actors involved in data analysis, from the purchase and sale of data as a commodity to its processing, make it clear that data minimization is very risky. UN agencies, aid organizations and academics usually require the anonymization of collected and shared data, no collection of data other than that explicitly stated, storage of data in a safe place, and destroying data when the time comes, but it is up for debate whether these can be monitored or produce the desired outcomes.

A wide range of techniques can be used to de-anonymize anonymized data (Véliz, 2020). Given practices such as collecting biometric data from people who apply for residence permits or visas or cross borders, taking their pictures, and asking for details of all aspects of their lives; seizing the phones of asylum seekers and examining the data on them, sometimes recovering deleted data; as was the case in some speculative projects, collecting genetic information via DNA tests; and, in the field of humanitarian aid, requiring the collection of identity information for mobile payments in order to make cash transfers, it is obvious that we are starting from a very weak point when it comes to data anonymization, as is also the case with regard to issues of privacy and consent. In the case of migrant and refugee movements, the problem is wider than the identification of individual migrants. Analysis of mass migration movements and their dynamics, independent of the identities of individual participants, can make it possible to take action against the movement as a whole, which in turn can have negative consequences for each participant.

In a data challenge in Turkey that involved analysing migrants' data sets that were shared by a phone operator, a team of researchers noticed that a group of migrants were working, informally, in the construction of a large infrastructure project. They chose not to include this finding in their report as, if it were made public, the migrants could lose their jobs. The awareness of this possible outcome actively prevented the negative consequences of revealing the information. This due diligence is not apparent in all instances of data analysis and, hence, similar circumstances are demonstratable to the potential harm to migrants under certain conditions in spite of the anonymization of data.

Consent, privacy, manipulation
Consent

The problem of informed consent, which is a source of great controversy and is encountered both in humanitarian aid and in AI ethics, partially stems from this new, market-oriented approach. In the field of humanitarian aid, when working with displaced people, one of the most important principles is to receive their consent, make sure that their participation is voluntary, and be careful not to force any decisions on them. This ethic connects to the no-harm principle. It requires implementing policies and providing aid only to the extent that recipients give their informed consent and offering alternative solutions otherwise. However, because technological and innovative solutions are required by donor countries, and large technology, finance and telecoms companies are involved in the process, both as new donors/funders and as active and opinionated stakeholders, decisions such as collecting data from migrants, digitizing the identification process and testing and implementing tech products such as blockchain and biometrics are made beforehand. Aid organizations then make proposals containing these requirements to get funding, following which a major infrastructure investment is made in the target region to facilitate the project, people are hired and, once everything is ready, the migrants are asked whether they give their consent.

The issue of informed consent was already controversial before the technological transformation, when the public-sector-oriented approach was dominant. A hierarchy had emerged between migrants who needed help and aid workers, and recipients had to accept the conditions attached and give their consent to access aid. Therefore, another approach called 'meaningful consent' was developed, which gave people a real choice and offered real alternatives. However, in the age of 'technology and innovation', the issue of consent is also becoming meaningless. After all this investment, and because there is no alternative and the focus is on ticking off the promised items within the given time frame rather than really improving the conditions, migrants and refugees find it impossible to understand the working principles of these technologies, recognize potential risks or make their decisions freely, and this is true for all new technologies from biometrics (HIP, 2020; Davies, 2021; Holloway et al, 2021) to data analysis to blockchain. Aid workers on the ground are also unable to comprehend the new digital technologies, provide detailed and clear explanations and convince people.

This issue also applies to those who have a critical perspective. In the Zataari camp in Jordan, for example, money was transferred to migrant people using blockchain technology, they had their eyes scanned to access their accounts when they went to grocery stores in the camp, and payments were withdrawn from their digital wallets (Juskalian, 2018). Critical academics and experts expressed a series of concerns, ranging from the use of biometric technologies (eye scanning) to blockchain, but it is usually not possible for migrant people to access this information, and they might only be concerned whether the scanning could damage their eyes and cause health problems.

In another example, migrant people in Nigeria were required to enter a passcode they were given in order to receive payments using their prepaid cards, but some were unable to do so because they could not read. The solution was a passcode that consisted of simple shapes, which gave rise to another problem: it was not clear which side of the card was the top, so people attempted upside-down shapes as the code. To solve this new problem, a human figure was drawn, with the head of the figure representing the upper side of the card. ... In short, in some cases, technology makes a process that should be simple more difficult and more complicated.

As a result, migrants and refugees end up accepting the technological and innovative solution forced upon them, whether they give actual consent. If they have particular concerns, they may choose not to be registered at the cost of not being able to access aid and having to deal with their problems on their own, or, as was seen in some examples where people had to provide their biometric data and have fingerprints taken, attempt to burn their fingertips to avoid detection by authorities later.

Privacy

Another related issue, of course, is privacy. Most of the examples in this section involve displaced people, but the disrespect of consent and privacy applies to legal migrants as well. Requesting mountains of documents and personal details from people who apply for residence permits or visas, as already discussed, implies that, if you would like to cross the border, be accepted into the refugee camp or need access to aid, you cannot raise any concerns regarding consent or privacy.

All of the surveillance technologies examined in this book are based on the rejection of privacy in the first place. Even countries that have strong privacy laws and privacy-sensitive public opinion experience serious problems with corporations having access to all sorts of information about people, categorizing them, predicting their attitudes in given situations and sending commercial or political messages accordingly and displaying posts that they are likely to enjoy. Surveillance capitalism rejects that people have privacy-related rights, and wants to access and analyse everyone's data. This is why Carissa Véliz (2020) argues in *Privacy is Power* that privacy is not a private matter between a corporation and an individual person, and it is intimately tied to relations of power prevailing in a society.

In the case of migrants and refugees, raising privacy objections results in the rejection of an application or denial of humanitarian aid. This is because monopolistic companies and security agencies are only concerned with accessing the data on people on the move and then work out which sector they will join as workers, and where and under what conditions they will become consumers and customers. This turns privacy into a disregarded and violated right.

A very interesting example that does not come up often in discussions is the European Space Agency's (ESA) activities in the field of migration and mobility. ESA's satellites have the power to identify a book read by a given person, and its methods are not new: they have been a part of our lives as the most important and effective surveillance tool since the 1960s. We tend to disregard satellites because we cannot see them and many of us do not know about their power. In recent years, ESA has been playing an active role in identifying migration movements and monitoring them from source country to destination country. Because of the nature of this sector, where human mobility is viewed as a risk factor by security agencies, it is quite understandable for ESA to have a close working relationship with military, intelligence and security agencies – it may be considered part of the security bureaucracy anyway. The concern is that ESA has been selling its data on migration. Companies, academic research groups and political organizations with different motivations can purchase satellite data from ESA and draw their own conclusions with the help of experts reading these satellite images. This turns it into an important source of risk. Irregular migrants or forced migrants might be able to observe security forces, robot dogs, drones, sensors and monitoring via smartphones or social media, but they are not very likely to detect satellites and take countermeasures (Bircan, 2022).

In fact, the sale of satellite images by ESA to anyone who asks, and the disregard of the possibility that people could be harmed as a result, show that we are living at a time when legal categories regarding migration have become meaningless and migrant people have stopped being viewed as human beings with rights: instead, they are viewed as members of a workforce whose movements should be monitored at all times, with various technological products being developed to this end.

Manipulation

One of the issues to be touched on in the form of short observations towards the end of this chapter is manipulation over social media. How migrants' social media posts, video views and search engine results are used to identify their goals and intentions and to present them with political and commercial messages fitting their profiles, and how authorities use this to manipulate, steer and control them, is an issue that, as far as I know, has not been studied in detail in the migration literature, and one that I would like to conduct a stand-alone study on. Revealed to the world with the Cambridge Analytics scandal, frequently discussed following the election of Trump in the US and the Brexit referendum in the UK, this issue eroded social trust in democratic institutions and mechanisms of participation and narrowed the public space, and, at a time when disinformation/misinformation and fake news have become a regular part of our daily life, social media has the potential to be used to manipulate migrants. Some of the well-known examples include Australia and the US buying YouTube ads to be shown in Afghanistan and Latin America respectively with the message 'Don't even think about coming to our country'. Alongside these, migrant people are also shown many ads about remittances, citizenship and language classes.

One concerning aspect of the issue is that commercial and political organizations have the potential to manipulate migrant people over social media. Another aspect, especially when it comes to asylum seekers on the borders, is that smartphone data and social media posts are considered to be reliable evidence by border agents. Migrants' phones are confiscated on the borders of the EU, UK and US (Riotta, 2020; Rohrlich, 2020), their social media posts and messages are examined and their photos are checked based on the belief that this will provide a reliable answer to the question of whether the applicant is a 'real' 'asylum seeker' or an 'economic migrant' trying to cross the border illegally.

Predicting migration

At the end of this chapter, I would like to give a few examples from the main objective of big-data analysis, which is basically predicting migration before it takes place. Collection of large data sets from real-time and historical sources and their analysis make it possible to predict mass migration movements and help to make decisions regarding individual applications.

An interesting, and by now classical example, is that the UNHCR uses goats to predict migration. People who intend to migrate from Somalia to Ethiopia sell their goats because they are valuable but difficult to bring along, and once they are in Ethiopia, they buy new goats. Once this observation was made, analysis of goat prices in the two countries made it possible to predict mass migration movements. A sudden decrease in goat prices in Somalia because of an excess of supply might indicate that a lot of people are preparing to migrate. Another consequence is that, in the later stages of migration, goat prices in Ethiopia would increase because of increased demand. This, in turn, allows authorities to prepare for mass migration movements.

Similar projection work is undertaken by the EU Agency for Asylum (EUAA), headquartered in Malta. Its goal is to detect irregular migration movements from Africa and the Middle East, and from Ukraine since 2022, and notify EU member states beforehand. In many cases, the agency has been said to make accurate predictions one month before a migration movement takes place. The EUAA also analyses its data in cooperation with many tech companies, data sources and security organizations.

In fact, big-data analysis forms the basis of the surveillance technologies discussed in this book and the argument regarding the trajectory of capitalism. The business model of surveillance capitalism depends on selling products and services to people by analysing as much big data as possible and predicting and manipulating the future. Intelligence plays a critical role in security bureaucracy's risk-oriented approach to migration and in detecting risks before they are realized, and big-data analysis forms the basis of this intelligence. On the other hand, smart border applications and digital identities (discussed Chapters 3 and 4), are also based on big-data analysis. Smart border applications make it possible to collect and analyse data and take action based on the decisions, and digital identities are products that collect and store data.

Moreover, the growing importance of big data and the motivation to collect as much data as possible from as many sources as possible result in the disregard of principles such as privacy, consent and no harm (developed on the basis of historical experiences), in the rejection of legal categories, and in the perception of all migrants as mere members of the workforce; while the desire to access new and different sources of data radically changes the position of humanitarian aid organizations with regard to monopolist companies. Private companies used to be part of the supply chains of these organizations, but now, aid organizations are becoming data suppliers for security agencies and companies, helping them to reach new markets, customers and data sources.

3

Smart Borders

Many of the technological solutions that stand out in the field of migration and border management fall within the scope of the tech product package known as 'smart borders'. These 'solutions' make it possible to control activities beyond physical borders and to detect and prevent immigrants before they even reach borders, through such tools as AI algorithms, unmanned aerial vehicles, facial-recognition systems, biometrics, satellite images, sensors and analyses of mobile phone and social media data. For example, Elbit, an Israeli security company that established an advanced surveillance system in Arizona, can detect people approaching the border from 7.5 miles away. The laser-enhanced cameras produced by Anduril, working in the same area, are able to detect all movement within two miles and distinguish human activity from animal activity through the use of artificial intelligence (Feldstein, 2019).

Smart border applications are the most visible examples when it comes to surveillance technologies in the field of border and migration management. The reason why I have discussed big-data analysis before moving on to smart border products is that smart border applications play a prominent role in collecting real-time data on human movements, and thus contribute to the development of surveillance technologies. Flows of real-time data collected using various methods including drones, satellites and sensors, and their analysis in comparison with historical data sets, play a very important role in predicting and preventing human movements and deciding who can be permitted to cross borders.

There are two basic issues involved, which have been discussed in Chapters 1 and 2 as well. The first is the risk that, if the advanced technologies developed here are applied successfully, they might be used to monitor and steer the daily lives of entire societies. I will discuss the pilot use of lie detectors from this perspective. The second issue is that these technologies are developed by military/arms firms, and the solutions they offer not only violate the basic rights of migrants and refugees by adopting a purely security-oriented approach, but they also militarize borders, contribute to intelligence work concerning other countries, put regional and global peace at risk and reinforce war preparations. They also provide ample opportunities for cooperation between tech companies on the one hand, and military/arms firms and security bureaucracies, on the other.

Academics and NGOs tend to have a critical attitude toward smart border applications. Of course, as discussed in this chapter, many academic research groups work with arms and tech companies in the name of research and development funds and pilot projects. However, they do not usually share their aims and findings with the public but conduct their work in relative secrecy. As a result, public discussion revolves around the more critical studies, which focus on specific regions, such as the Mediterranean or the US-Mexico border, and on specific types of migration. In particular, these technologies are known to be used actively in detecting and preventing movements of irregular migrants and refugees from the Middle East and north Africa into Europe, and from Mexico to the US. Drones, sensors, social-media analysis and robots, among other methods, are used to identify people moving towards borders so that they can be prevented or pushed back, or to inform security forces in Libya, Turkey or Mexico so that they can intervene. As a result, migrant people are forced to end up in countries other than their preferred destinations where they would like to take refuge.

Different consequences of smart borders in different regions

Smart border applications do not always generate the same results. For example, the same products that have lethal consequences on the borders between Turkey and Greece, Mexico and the US, and Italy and Malta with Libya, can make life easier when deployed on borders between certain other countries. For example, the thousand-mile land border between Norway and Sweden, the longest in Western Europe, is frictionless thanks to the cutting-edge border technologies used. It is possible to scan vehicles for prohibited or restricted goods that need to be declared to the customs authorities or monitor vehicles crossing the border using automatic number plate recognition (ANPR). This makes it possible for about 70 per cent of freight trucks to cross the border without stopping and saves a lot of time for people who work or have business in the two countries, without wasting resources on passport control or customs inspections.

Something similar might be possible for legal migrants as well. As discussed in Chapter 2, people who apply for tourist, business or student visas, or for residence or work permits, are asked to provide information on almost all aspects of their lives, which are then analysed. Many countries, such as Canada, are considering the automation and streamlining of visa and permit processes using AI applications (Akhmetova and Harris, 2021). Therefore, if you are not a citizen of a country classified as 'risky', and if you have the required documents, it would be possible to obtain your visa or residence permit without much delay. Moreover, various AI algorithms are being developed to help case workers or border agents with refugee applications in some EU countries, for example in Germany. The ultimate decision is made by a human being, not by these algorithms, but they help officials with several issues such as analysis of the information submitted by the applicant or the transliteration of Arabic names to the Latin alphabet (Bankston, 2021).

When it comes to smart borders, these technologies are being used in diverse fields and under many different conditions, and the same products can be observed to generate opposite results in different regions or under different conditions. A product can lead to the deaths of thousands of migrants in the Mediterranean when used by Malta, or mean that migrants end up in slave camps in Libya, but can facilitate and streamline trade and human mobility in Norway.

However, these technologies are developed by the same companies in both fields, with similar goals. This, in turn, offers clues as to the direction of contemporary capitalism and shows how surveillance capitalism has become a structural feature of it. This point was made in Chapter 1, and is partly about speeding up the free movement of goods and services, and, for example, increasing trade by streamlining border crossings by trucks, and partly about reshaping and repositioning the working class in terms of migration and human mobility, and deciding who will be allowed to be mobile, up to what point, and who will be included in the capitalist processes of production.

Capital, goods and services can circulate around the world at breakneck speed, but labour cannot move as fast. However, new technologies make it possible to manage, direct and/or prevent mass migrations and human movements without employing thousands of people for migration and border management. So, the same technologies may generate different results for migrant people in Norway and Greece, but this cannot be attributed to Norway being a nice, pro-migrant or hospitable place. The ultimate cause is the interests of capitalism and the process of shaping class relations.

Another aspect of this issue is that technological investment in border and migration management provides important examples regarding the deepening of surveillance capitalism and increased control over, management of and – as a natural result – power to manipulate entire societies. As the Cold War made abundantly clear, the military and security industries in advanced capitalist countries play a dominant role in the development of cutting-edge technologies. Thanks to heavy public funding and long-term research opportunities, many of the technologies that we use today were first developed under secret conditions for military and intelligence purposes. The success of Silicon Valley and the emergence of a large number of civilian tech companies gave rise to the argument that technological production is now democratized, and we are no longer dependent on the military industrial complex for its development. However, global tech companies, not least those headquartered in Silicon Valley, come together with military and security agencies and undertake joint projects. Border and migration management proved to be a fertile field for such cooperation to flourish (Larsson, 2020). Of course, there is ongoing cooperation to develop weapons and military technologies as well, but it still takes place in relative secrecy and is only noticed when some of these weapons are tested in actual combat; whereas, in the case of border management, a much larger number of products are launched in shorter periods of time and are installed on the borders and tested by states all the time.

One of the most important features of surveillance technologies installed on borders, which facilitate human movements in some places and prevent them in others, is that they speed up the development of technologies that collect and analyse real-time data. Tracking and analysis of mobility using real-time data, apart from categorization and probability calculation work conducted using historical data sets, play a key role in the development and deepening of surveillance capitalism. The business model of surveillance capitalism was developed thanks to offering customized ads to users based on analysis of real-time data, which Google achieves using its search engine and map application, Amazon achieves by analysing shopping patterns and viewing habits, and Meta through analysis of social media posts. That this went beyond being a simple business model and became a determining feature of contemporary capitalism is made clear in the case of controlling mass movements of people.

One aspect of the technological advances in this field is visible in systems such as facial recognition and mobility pattern analysis, which are used by metropolitan police agencies for crime prevention and predictive policing purposes (Milivojevic, 2021). They are also used to predict and prevent mass protests and uprisings. The other field where the advances made their presence felt is the field of border and migration management. As a result, tech products claimed to be developed for border management are purchased by authoritarian governments, such as that in Uganda, to monitor and suppress local opposition. This is because the most important capability provided by surveillance technologies is to protect political power through social control and manipulation. Just as the weaknesses, needs and tendencies of people identified using social media and digital platforms are translated into a means to manipulate in order to sell commercial products or political messages, strengthening the surveillance aspect of these technologies is critical for social control and manipulation.

At this point, we have much to gain from adopting a perspective framed by capitalism because the problem is not limited to oppression by authoritarian governments in the Global South; rather, the problem is also that societies in advanced capitalist and democratic countries with strong welfare states, such as Norway and Sweden, are enticed into being consumers and producers in line with the needs of capitalism. The sale of these tech products by advanced capitalist countries to less-developed countries allows the governments of the latter to remain in power, and, more importantly, reinforces the dominance of imperialist/capitalist states over dependent countries and strengthens neo-colonial practices. It offers opportunities both to control the collected data using the products sold via technology, finance and military/intelligence companies, and to deepen political dependency.

The development and testing of smart border technologies

Examining smart border technologies used in migration and border management is critical for understanding the scope of the surveillance technologies being developed. Although there is a certain amount of secrecy surrounding this issue, product testing and pilot projects in border and migration management provide an idea of their extent.

There are two examples that first come to mind. The first is how Israeli arms and security firms market their surveillance tech for border management as being 'successfully tested' on Palestinians. Many products are first used to monitor and prevent the protests of Palestinian people, mainly in Gaza, and this is used to demonstrate the effectiveness of the products to potential customers (even though the firms' claims are open to discussion). In this sense, Gaza serves as an open-air prison, is kept under continuous surveillance and used for testing these technologies.

The second example is the high-tech 'new generation refugee camps' established on two Greek islands in the Aegean in 2021 with the financial support of the EU (the EU paid €43 million for the camp on Samos, for example) (Molnar, 2021b). These have attracted attention for their prison-like structures and the miscellaneous tech products installed. Biometric recordings and the cameras and sensors installed all over the camps allow the thousands of asylum seekers in the camps to be monitored constantly, with every movement analysed using AI algorithms, and all potential security threats reported to the Control Centre in Athens (Molnar, 2021a). In other words, these camps are violating the right to asylum, personal liberty, and private life, and are another demonstration of new technologies being tested on refugees.

Both the US and the EU have hired private companies to militarize their borders with advanced surveillance technologies. In other words, public funding is not being used here to ensure the rights of immigrant people or to resolve economic and political problems, but to support companies in the development of new and deadly technologies. For example, the border control and immigration budget of the US has increased by 6,000 per cent since 1980, with US$223 million of the 2019 budget allocated to Homeland Security for the development of border security technologies. In 2018, Northrop Grumman signed a US$95 million contract to establish a biometric database for Homeland Security, and the collected data is currently being stored in the Amazon Web Services cloud. Meanwhile, the EU has allocated €34.9 billion to this field for the 2021–2027 period, up from €13 billion for the 2014–2020 period (Daniels, 2018; Feldstein, 2019; Achiume, 2020; Sánchez-Monedero and Dencik, 2020).

Frontex, which operates on the land and sea borders preventing the movements of migrants in Europe, uses the most advanced surveillance technologies and conducts push-back operations, barring immigrants from the opportunity to seek asylum. In its operations, Frontex uses military-grade drones, and has signed a €50 million contract with Airbus for aerial surveillance purposes (Achiume, 2020; People & Planet, 2021). Accordingly, people on the move are forced to access their target country via more dangerous routes, meaning that a greater number of people die on the road.

The UK is another country that invests heavily in surveillance technologies. According to a Privacy International report (2021), ADS – the UK's main arms lobby group – has established an Industry Reference Group together with the Home Office to lead the process. Within this scope, Future Borders and Immigration Systems, established in 2020, has been assigned a budget of £113 million for the development of digital borders. UK arms company BAE Systems, and CACI, a US security company, both won government tenders, in 2018 and 2020, respectively, each valued at £4.9 million. In January 2021, Tekever, a Portuguese weapons company, was given a budget of €6 billion to use drones to detect boats carrying migrants on the English Channel. Finally, making use of software developed in 2018, Cellebrite, an Israeli surveillance company, can analyse all available and deleted data on the phones of migrant people and scrutinize any suspicious activity on migration routes (Privacy International, 2021).

The US encourages these developments too. The RAVEn program, for instance, developed by the DHS at a cost of US$300 million uses AI algorithms to identify non-US citizens via social media mining, biometric data and location-related data and detect suspicious cases (Mijente, 2022). Border security is not limited to contracts by border security agencies. For example, in 2019, the US Marine Corps awarded a US$13.5 million contract to Anduril to develop a system to automatically detect, identify and classify all human and vehicle movements along a 30-mile stretch of the US–Mexico border. In the US, the infrastructure for and investment in smart border applications began under the Obama administration. Under Trump, use of digital products on borders expanded with the participation of Silicon Valley firms. This trend continues under the Biden administration. Clearly, attributing such developments to Trump's anti-immigrant views would not be sufficient, as the cooperation between US government agencies and technology/military companies continues apace. These companies include Palantir, Amazon, Salesforce, Microsoft, Dell and Hewlett-Packard, among others (Mijente, 2019).

Facial-recognition technologies, which give rise to the gravest risks in terms of mass surveillance, are another popular tech 'solution' in border and migration management, viewed as just another smart border application. Its use is banned in some US states, but police forces in other states and in many countries use these technologies in situations ranging from passport renewals to border crossings. Facial recognition based on people's digital still and video images can be used for identification, verification, categorization and characterization purposes. It is a nascent technological field, has a higher rate of identification failure among people from minority ethnic groups, and its use by security forces is very risky because of its low accuracy rates. Nonetheless, the global market for facial recognition is expected to grow to US$12 billion by 2028, and migrant people are used as test objects to perfect the technology and reach this market goal (Kelly, 2022).

The most significant producers of smart border applications are military companies. So, those who produce the weapons and bombs that force refugees to flee their countries are the same companies that produce the tools for detecting those people in border areas. For example, leading weapons producers such as Lockheed Martin, Airbus, Safran and Thales and tech companies such as IBM, Amazon, Microsoft, Fujitsu and Accentura all stand out with their investments in this area (Achiume, 2020; Au, 2021).

A side note on Palantir

The World Food Programme drew criticism for sharing data on more than 92 million beneficiaries with Palantir in a US$45 million deal (Achiume, 2020). Founded in 2003, Palantir works with US intelligence, military and law-enforcement agencies in many fields, including during the wars in Iraq and Afghanistan. The company received start-up funding from In-Q-Tel, the venture capital arm of the CIA. It is no secret that Peter Thiel, the founder of the company and a former advisor to Donald Trump, has had influence in the past over political decision-making (Barbrook and Cameron, 1996; Cohen, 2017). Many former Palantir employees work for the Pentagon, and Palantir employs a large number of Pentagon veterans. From 2006 to 2019, Palantir spent up to US$15 million on lobbying activities and was handsomely rewarded with big contracts. For example, in spring 2019, it was awarded a US$800 million contract for setting up a real-time intelligence database for the US military, overtaking a conventional military company that had been operating in this field for many years. Palantir was awarded contracts worth more than US$150 million by Immigration and Customs Enforcement, developing, for example, the case-management tool that separated parents from their children in families crossing the border together, which cost US$53 million in its initial phase. When other agencies such as the FBI and the US Navy are included, the company received contracts worth more than US$1.5 billion in total (M. Burns, 2015; Hatmaker, 2019; Mijente, 2019).

Palantir has faced much criticism, aided perhaps by the caustic personality of Thiel, but other tech companies are hardly different, intertwined in many similar close relationships with security services. At the end of the day, Palantir uses Amazon's cloud services to run its programs. Amazon and Microsoft competed with one another for a US$10 billion contract to build a unified cloud computing system for the US Department of Defense. Eventually, Amazon was awarded the contract, making it the biggest military industry company in the US. Microsoft received US$480 million to build a virtual-reality system for the US military. Project Maven, an AI drone-targeting program, was first awarded to Google, but when Google employees protested, the contract was given to Anduril. Google employees used the company motto 'doing good' in their protests, and Google have since stopped using the motto. Anduril is a company that has set up surveillance towers for the UK Royal Marines, US Marine Corps and now, the US Customs and Border Protection (Mijente, 2019).

Speculative academic projects

Miscellaneous speculative pilot studies are being conducted in the fields of migration, refugees and border management. For example, in a project called Avatar (Automated Virtual Agent for Truth Assessments in Real Time) applied in Canada, the US and the EU, an AI lie detector algorithm has been tested during which people answered questions asked by kiosk computers at border crossings and airports. The algorithm tried to detect if the person was lying or not by analysing their eye, mouth and hand movements, using sensors, biometrics and facial-recognition systems (Daniels, 2018).

These projects are usually pilot studies, and so are supported by public funds in the guise of academic research. In the UK, for example, another lie detector algorithm was developed by QinetiQ, a British defence technology company funded by the UK Engineering and Physical Sciences Research Council. Another example is the iBorderCtrl project conducted by Manchester Metropolitan University researchers, which benefited from a €4.5 million grant from the European Research Council within the scope of the Horizon 2020 programme (Sánchez-Monedero and Dencik, 2020). This project, that makes use of affect-recognition technologies, was tested in Greece, Hungary and Latvia (Feldstein, 2019), while another speculative and highly criticized pilot study is the UK Border Agency's Human Provenance Pilot project, which aims to detect the nationality of asylum seekers through a DNA and iSTOP analysis (Akkerman, 2021).

Tech or military corporation or both?

How should technology companies that are a part of surveillance capitalism be defined? Can such companies as IBM, Microsoft, Amazon, Palantir and Google be described as mere technology providers? For example, Amazon, Microsoft and IBM are in a race to provide cloud services and digital infrastructures to many military and intelligence organizations, including the US. Ministry of Defense and the CIA (Feuer, 2020). They are active in the development of many state-of-the-art combat technologies, such as autonomous weapons and robot soldiers. What is more, they have people who have retired from or been transferred from state security institutions such as the Pentagon and CIA on their boards of directors and in senior management. These companies are operating actively at a state level, providing lobbying and consultancy services and assuming various official positions while enjoying public funds. As such, they go hand in hand and have mutual interests with the military/security bureaucracy.

When the security bureaucracy buys the products of tech companies as a customer, or when these companies develop products with public institutions using public resources, border security and the prevention of the movement of migrants and refugees become a fruitful area for such collaboration. That is when military companies that traditionally work in this sector will come into play. Like finance, the defence sector is an area where advanced technologies are developed and applied. Historically speaking, technological developments in the military sector have been a result of the efforts of public companies and a few public-private partnerships, although in recent years, 'civilian'-private companies have become prominent, especially in the fields of cloud systems, sensors, satellite-space technologies, machine learning and robotics, and it is these companies that are reintegrating with the military sector and bureaucracy (Larsson, 2020). In line with Lenin's (2008 [1917]) definition of imperialism – the integration of industry and financial capital in an oligarchical structure – today's global technology companies also form part of this oligarchical structure. Since it is now financial and military companies that develop and apply advanced technologies, there is no way of distinguishing between tech companies and military and finance companies, and their relationship with public agencies and their personnel serves to consolidate this integration.

Alongside the militarization of borders, the field of immigration and border security is also a fruitful area for the transformation of surveillance capitalism from a business model to a structural and critical feature of the capitalist system, and for the effective use of surveillance and manipulation on society through advanced technologies. As such, migration management and border security accelerate the cooperation and integration of these companies with bureaucratic bodies, allowing the most advanced and cutting-edge technological 'solutions' for the control and manipulation of society to be tested. Today, the prevalence of migrant and refugee prejudices is facilitating this integration and opening doors for the violation of rights and speculative tests in a more reckless fashion.

Smart borders and threats to peace

We are talking about a cooperation process that involves large investments and massive profits. The shaping of human movements in line with the interests of capitalism affects all structures of capitalism as a whole, apart from the core oligarchical structure. As argued in Chapter 1, this process benefits industries that rely on migrant labour in their supply chains, from textile and apparel firms to food companies. However, the reinforcement of suppression and control mechanisms are not limited to the manipulation of societies. They also increase the risk of war and make it more difficult for disputes among countries to be resolved in a peaceful and diplomatic manner. The critical importance of drones has become all too apparent in the Armenia-Azerbaijan war and the Russia-Ukraine war, as well as in the so-called anti-terrorism operations undertaken by various countries. As dancing robots start performing sentry duties and taking shooting practice, we will soon be seeing robots on the battlefield as well. Cyberattacks and cyber-weapons can destroy or sabotage the infrastructure and communication network of the enemy and create panic and chaos behind the front lines (Cummings et al, 2018; Kello, 2019). Technology also makes it possible to carry out complicated intelligence operations and inflict heavy damage as in the attack on Iran's nuclear facilities, the assassination of an Iranian general using an AI-operated weapon, or the Pegasus attack (Cottray and Larrauri, 2017). This is in a world where states have experience in controlling and steering their own societies, carry out round-the-clock surveillance beyond their borders in the name of migration management (usually in the territory of another country), collect data from citizens of other countries under the name of visa applications, and can monitor and intervene in political and economic processes in other countries via social media companies and digital platforms. In a world where food security and access to basic resources have become critical issues because of climate change and the increasing economic and technological dependency between a small number of capitalist countries and others, efforts to solve conflicts via peaceful means have been weakened.

It has been widely established that capitalism has a tendency to cause wars over the distribution of markets and resources (Lenin, 2008 [1917]). A factor that strengthens this tendency is that surveillance technologies developed by the oligarchical structure in question have the capability to be turned into weapons of war very quickly, increasing the probability of local, regional and international conflicts (Akkerman, 2016; Larsson, 2020).

A related issue is that an aggressive attitude adopted by actors who consider themselves to be stronger because of their technological superiority may backfire. The increasing virtualization of borders strengthens border security in some respects, making it possible to monitor larger areas more effectively and with fewer personnel. However, every virtual system is susceptible to system errors, crashing and hacking. Investment in smart border applications in the name of stronger security may, paradoxically, render borders more vulnerable and permeable. Security vulnerabilities could arise because of the hacking of virtual walls and borders or a systemic error, leaving the country in question open to attacks (Kello, 2019). Moreover, inaccurate or deficient analyses of human movements from a cybersecurity perspective may create new vulnerabilities on the part of migrant people. Because of the complicated nature of many of these technologies, detecting and solving problems rapidly or coming up with alternative solutions might be difficult.

In addition, because these technologies are developed by a handful of companies based in a small number of countries, there would be consequences when states that have not produced or developed these technologies use them to protect their borders or undertake military operations. The interests of or interventions by companies that hold the licenses or patents for the products and have better control over their working principles, as well as the military/security bureaucracies with which they are in close contact, might have repercussions for the scope and direction of military conflicts.

Should we aim for reform in opposing surveillance technologies?

There is an extensive literature on racial capitalism. New technologies are being used to reproduce the dehumanizing attitude that imperialist capitalism has adopted towards societies in its historical colonies. The racial hierarchies that existed in the minds of the people who legitimized slavery in the early ages of capitalism have permeated the system that develops surveillance technologies today. As a result, algorithms based on deep learning make decisions that align with white supremacist and imperialist goals. The violence, discipline and control that colonized societies have been subjected to systematically in the past are now being implemented more effectively thanks to the technologies collectively referred to as smart border applications, from a distance, in a 'sterile' and 'hygienic' manner for those who wield them, and without having to face the people subjected to violence. Therefore, it is not realistic to expect to reform this system based on human rights via transparency and better supervision in the processes of machine learning and coding. In practical terms, it is not even possible to track the exact steps in the machine learning or deep learning process, understand what happens in the black box and make interventions (Gilroy, 1993; Mbembe, 2003; Beniger, 2009; Benjamin, 2016; Gazzotti, 2021; Nonnecke and Dawson, 2022).

Schulz (2022) criticizes the critical approaches to smart border products taken in academia and civil society, arguing that offering to solve problems by adopting a human-rights-based approach and aiming to reform the use of technology in border management is a techno-liberal approach and cannot create real change because it does not question the system that creates the problem. She also discusses the risks that would arise with the automation of an already biased and unjust system and an abolitionist approach that proposes to construct a new institutional structure. She recommends re-designing the entire process with a migrant-oriented approach.

It is clear that the issue is not limited to migration: a comprehensive, long-term and militarist approach uses operations involving migrant people to create and test new military structures. Frontex, for instance, has been facing criticism since 2020. Its executive director Fabrice Leggeri resigned in 2022 when it was documented that Frontex violated the rights of asylum-seekers by leaving them to die, giving push-back directions to Greek authorities by providing them with coordinates, and advising the Libyan coast patrol so that they could pick up asylum seekers. In statements made to the public and the Frontex personnel, Leggeri had defended the activities of the organization, arguing that Frontex was a security organization, not a "welcoming committee", and had to take tough measures against people attempting to cross borders without authorization. The director, who transformed Frontex from a small agency into a border-security organization that operates on all borders of Europe, employs thousands of armed and uniformed personnel, and works in coordination with the border-security agencies of member states, apparently had a vision to turn Frontex into the core of a future European army (Bird, 2022; Lighthouse Reports, 2022).

While this book has generally been focused on the socio-economic direction of capitalism, the preceding section has demonstrated the importance of literature regarding racial capitalism. Note that the socio-economic direction of capitalism lies at the root of racial and colonial approaches – also working to reinforce one another. At this point, we can look at two examples of the effective use of surveillance technologies in the reproduction of racial and colonial approaches. The first involves the use of drone and satellite images to detect people attempting to cross from north Africa into Malta and Italy, and then informing the Libyan 'coast guard' organization – which is controlled by one of the warlords fighting in Libya – even when the boats are in international waters or in Malta's territorial waters. The warlord's gang then takes migrant people from EU waters and puts them in slave camps in Libya. Despite all the documents and migrants' statements confirming the practice, no international organization, including EU and UN agencies, have taken any steps to counter this activity; on the contrary, they continue to provide financial support to warlords in Libya.

The new generation, high-tech refugee camps established on Greek islands, mentioned earlier, constitute a more blatant example of racial discrimination and colonialist attitudes. In these camps, hundreds of refugee people imprisoned in a laboratory-like environment are monitored around the clock. Cameras, biometric records and sensors keep generating real-time data on them. People put in these camps are asylum seekers from Africa, the Middle East and South Asia. The camps enable testing of many new tech products. For example, the spread of these and similar camps would offer an important opportunity to tech companies working on facial recognition and emotional AI. There are many reports that facial-recognition systems are not as accurate when it comes to Black people and people from minority groups and ethnic communities, resulting in discrimination. Similarly, there are important limitations arising from ethical and privacy concerns when developing emotional AI systems. These labs disguised as refugee camps offer opportunities to surveillance capitalism in risky but 'promising' fields.

Returning to the question of reform or revolution, we need to ask whether our main concern is creating better and more accurate facial recognition systems. Are we trying to make sure that these systems are just as accurate for people from all ethnic backgrounds, or should we oppose the use of facial recognition systems by military, intelligence and security organizations, given their repercussions for the whole of humanity?

Smart borders and the rejection of privacy

Another consequence of the racial and colonial approach is that issues such as privacy, which are very important for societies in advanced capitalist countries, are ignored or denied when it comes to migrants, refugees or people of historically colonized countries. For example, it is common practice for authorities to seize the phones of asylum applicants and examine their contents without the consent of their owners. Smart border products that support these practices include those that restore data deleted from smartphones, which are also very popular. When asylum applicants have happy images on their phones, such as pictures of wedding parties or festivities with their friends and family, this is used as evidence that they are not actually fleeing war or persecution. When the pictures are deleted, this is then interpreted as admitting as much, and the applicant is considered to pose a bigger risk. Attempts to avoid putting one's travel companions at risk or to hide one's ties to people smugglers are sometimes viewed as aiding and abetting (despite the fact that you must cross borders illegally to be a refugee or asylum seeker, which in many countries requires working with smugglers). Similarly, there is no contradiction between people wanting to preserve good memories of their families, loved ones and hometowns, and fleeing war, persecution, violence and poverty.

Practices such as seizing phones, using lie detectors, monitoring and listening in on people even before they approach the border, using fake phone towers and various sensors, are other indicators of a denial of privacy, but states view these practices as their right to ensure their security as sovereign actors.

However, the argument that such privacy-eroding practices are essential to maintain the sovereign rights of states does not hold up well to scrutiny. Surveillance technologies allow states to detect migrants long before the moment they reach the border. For example, there are many efforts in the EU to use digital data from satellite images, social media and smartphones to predict whether people departing from central Africa have Europe as their ultimate destination, while they are still in central Africa. They resort to the power of data analysis to distinguish migration into Europe from intra-African migration, which is much more common than on other continents.

This is a component of the common phenomena of border externalization, which refers to the practice of sharing conclusions drawn from data analysis with transit countries so that they can stop the migration movement in question. Agreements that the EU signs with north African countries as part of its neighbourhood policy (Deridder et al, 2020) contain provisions for the prevention of migration and transfer of border-security technologies to these countries to this end. Therefore, sending undesired arrivals in Europe back to their countries of origin – which we have already discussed with many examples – is not the only goal; another is to make sure that countries such as Morocco and Tunisia serve as buffer zones to prevent migration into Europe (Andersson, 2022). A similar case is that of Turkey, which agreed to prevent migrants' crossings with the readmission agreement it has signed with the EU. It is possible to examine the historical, socio-economic and cultural conditions that give rise to these policies of border externalization, but, importantly to this current book, it is clear that surveillance technologies play a critical role in their effective implementation.

The most problematic examples of new technologies in border and migration management concern the prevention of irregular border crossings and asylum applications. These examples naturally attract a lot of attention, but it is important to keep in mind that these technologies are used on all people on the move. Practices such as slavery camps and high-tech refugee camps resembling prisons and laboratories, briefly discussed earlier, are blatant examples of colonialist and racist approaches that remind people from former colonies that they are not viewed as full human beings. However, people who attempt to cross borders on boats or hiding in trucks make up only a small percentage of all migrants. They constitute a small percentage of irregular migrants as well. Many migrant people go to their destination country with legal visas and become irregular migrants when they overstay these visas. This is part of the reason why, in the case of migration from Africa or South Asia to Europe and from Latin America to the US, legal migrants are also subjected to surveillance technologies, and people continue to be monitored even when their stay is legal.

This shows that, even in the presence of strong privacy laws in a destination country, migrant people's right to privacy is dismissed outright or respected only to a lesser extent. This is also clear from the practice of requiring visa applicants to submit documents on all aspects of their lives. Moreover, collecting biometric data has put down the technological foundation for continuous monitoring. The requirement to provide new information on developments in one's life for visa renewals, as well as digital identity applications for financial integration and other purposes (discussed in Chapter 4), results in migrants giving up their privacy rights in order to obtain residence permits and, after a while, citizenship in their new countries. Moreover, given practices that violate the privacy rights of an entire society and abuse them for commercial and political interests, despite the presence of strong privacy laws – as revealed by privacy-oriented NGOs, academics and journalists – migrants are sometimes forced by states to give up their privacy rights regardless of explicit state policies.

Smart borders and denial of the right to asylum

Chapter 1 discussed capitalism's approach to migration and argued that legal categories of migration have lost their meaning, refugees and asylum seekers are denied their rights in practice and surveillance technologies have helped to create an environment in which the main determinant is the commercial interests of capitalism, and the reshaping and repositioning of the working class have become the main concerns. The role played by smart border applications in the de facto denial of the right to asylum deserves a closer look.

As mentioned, you cannot apply for refugee or asylum-seeker status in your own country. You first must cross the border using your own means, and lodge your application, in person, with the authorities of the other country. Therefore, crossing borders without permission or receiving support from people smugglers is not a crime for refugees and asylum seekers. (Although, if your asylum application is rejected, you may face deportation or a lengthy legal process.) This right is formally recognized but denied in practice by many countries. For example, the UK planning to send asylum seekers back to the safe third country they arrived from or to Rwanda. The argument is that refugees should stay in the first safe country they arrive in and make their applications there. In legal terms, however, there is no such requirement. Refugee people may cross multiple countries for various reasons and apply for asylum in their destination country. Moreover, this idea does not sit well with neighbouring countries that face large inflows of refugees and asylum seekers. That Turkey, Greece and Italy should have large refugee populations while the UK or Sweden reject applications on the pretext of a 'safe neighbouring country' is considered to be unfair, and there are loud calls for burden-sharing when it comes to refugees. The most common means of burden-sharing is to provide financial support to host countries, and the second is the resettlement of a certain number of refugees every year, which the EU arranges internally within the framework of the Dublin Regulation and the US does through the promises it makes to UN agencies. Many resettlement programmes work on this basis. In practice, however, resettlement programmes are far from meeting the demand, and countries do not even keep their promises regarding the number of refugees to admit.

Smart border applications play an important role in the denial of the right to asylum and the prevention of new arrivals. One aspect of this is the use of drones, satellites and sensors to detect people who approach the border. Another aspect is the collection of data against people who somehow do manage to lodge an application, by inspecting the contents of their phones or using lie detectors, among other means.

Resilience

Another range of technological products is in the innovative development work undertaken in less-developed countries by advanced capitalist countries – which deny the right to asylum in practice or avoid implementing it – donning their donor-country hat and in cooperation with UN agencies and humanitarian aid organizations. Of course, the classical liberal argument that every country will one day be 'developed' is an empty one given the worldwide division of labour in the capitalist system, disputes over market access and different levels of development and dependency relations. Moreover, in the case of displaced people fleeing internally or from countries ravaged and destroyed by wars and climate crises, it is unrealistic to expect development projects funded by donors and undertaken in neighbouring countries – which have to deal with hundreds of thousands if not millions of people seeking asylum at once and already face similar economic, political and environmental problems – to be successful.

The concept developed in the face of this reality is resilience. Resilience aims for people to recognize and work to improve their conditions and strengthen their ability to persevere in adversity. These humanitarian/

donor actions are preventative of migration – indeed, the logic of resilience is imposed to prevent migratory flows towards the donor states –another example of border externalization (Bigo and Guild, 2005; Bigo, 2007).

People may have become displaced out of fear for their lives, but economic, social and cultural demands take precedence once they arrive at their destination, and they want to have access to services such as education and healthcare and participate in the labour force. These are not easy problems to solve. Donor countries place special emphasis on innovation as they provide support to resilience-oriented projects run in cooperation with aid organizations and UN agencies, which are the ones that undertake activities on the ground and work with people face to face. For projects to be innovative, they must use a certain level of technology, and this requires people to have, at a minimum, a (smart) phone and access to the internet or a mobile network.

For example, it might be a good idea, in terms of increasing resilience, to offer cash transfers instead of in-kind support so that people would shop in local markets, meeting their needs while supporting local producers. Making the cash transfers via mobile payments makes sense in terms of oversight and security, while adding to the innovative aspect of the project. It also means that banks, phone operators and tech companies must be involved. As a result, a significant portion of the donor funds end up being paid to these companies, which also acquire new customers and gain access to data from large groups of people who were previously inaccessible. Or, to increase people's resilience, they might be hired to perform remote work for companies in other regions because there are not many job opportunities in their places of residence. Coding lessons or digital literacy training might be lectured to this end, or smartphones, SIM cards, computers and internet access might be provided so that people can take up remote work. For example, hundreds of displaced people in Africa and the Middle East work on machine learning projects for Big Tech companies. Indeed, the biggest 'AI factory' is not in the US, it is located in a camp in central Africa. Every day, during their shifts, employees show pictures to computers and reject or accept the responses provided by the algorithm. It is very costly to hire students – let alone the rest of the population – in the US or UK to perform these simple tasks, but quite profitable to employ refugees and displaced people in various African and Middle Eastern countries such as Jordan and Lebanon, with the added benefit of great PR.

In another example concerning resilience and employment opportunities, donors and aid organizations consider developing the gig economy and digital platforms when the country in question has little capacity to create new jobs. For example, many projects have been proposed in Jordan since 2015 to ensure the participation of refugees in the labour force, but nearly all have failed. One project aimed to employ refugees in textile factories established in the middle of the desert, in so-called special economic zones, with the ultimate goal of exporting the products. However, people preferred informal jobs in cities to working in the desert. When similar initiatives failed to generate the desired results, a new effort began, in 2018, to make sure refugees earn some income by working freelance jobs through the gig economy. Unlike other examples in many parts of the world, it is the humanitarian aid bodies that support start-ups and provide financial and mentoring services to them. As a result, various digital companies such as Uber started operating in these countries. In this approach, men work in jobs like taxi driving, whereas women, complying to societal gender roles, are offered roles such as sewing garments at home, working as cleaners in other homes, and cooking meals for delivery. In this example, refugee people may be employed by different digital companies operating in textile, catering, cleaning and delivery sectors.

In all these examples, the success of a project is measured by the employment of individuals with support from humanitarian organizations funded by donor countries, engaging in production, making purchases in the market, and the revitalization of local economy as a result. This can also be quantified by looking at how many people are employed and how much income is generated. These projects may have relatively improved the conditions of many people, but whether they increase resilience is a matter of debate. Moreover, there is no agenda for increasing social resilience in advanced capitalist countries, which is only preached to people fleeing wars and climate crisis in the Global South, another indication of the colonial and racial approach. Making more money does not necessarily pull people out of poverty, and they may still have educational, health-related, cultural and environmental problems.

The success and the contents of the concept of resilience can be criticized from this perspective. As these examples show, a market-oriented approach ends up expanding the sphere of capitalist influence, helping many global companies invest in new countries, and making hundreds of thousands of people work in the supply chains of capitalist companies and start purchasing their products. People are led on with fake promises, and their weak positions are abused while capitalism ends up the real winner.

This chapter has focused mainly on smart border products and has discussed the control, monitoring and prevention of migrants using border security technologies. Surveillance technologies on borders adopt an adversarial attitude toward migrants, question them, analyse their data with suspicion and, in so doing, reproduce racial and colonial practices. The last section has discussed projects that are claimed to benefit refugees and displaced populations by supporting their livelihoods and helping them with employment. These projects are funded and products are made by the same companies and states. Thus, in the case of surveillance technologies,

I believe we would be well advised to go beyond criticizing anti-immigrant approaches and train a critical eye on apparently pro-migrant activities as well. The discussion on digital identities in Chapter 4 will show how surveillance capitalism gained dominance by adopting a seemingly pro-migrant approach, also making the connection between smart borders and digital identity initiatives.

4

Digital Identity and Surveillance Capitalism

This book examines the role of migration and border management in understanding the process by which surveillance capitalism has become an intrinsic and fundamental component of contemporary capitalism, and discusses the reasons why many technological products are being tested on people on the move, as well as the potential ramifications. Up to this point, I have first discussed the concept of capitalism, followed by a discussion of its consequences in the fields of big-data analysis and smart borders. Now, I will take up a more complicated field. I will consider efforts for digital identity, or the digitalization of identity, in migration and border management.

The main reason why this is more complicated is that goals such as supporting migrants and refugees and contributing to their financial and social integration are at the forefront. Data analysis and smart borders, in particular, function to identify, track and, when necessary, stop people on the move, and decide and limit how far they are allowed to go. Basically, they are fields in which states and certain industries, such as border management and war technologies cooperate, with technology companies. Here, the lines are clearer. There might be people who view these tech products positively for reasons such as preventing migration movements, reinforcing national security and border security and fighting smugglers. Or, the 'security-oriented' approach in question may be rejected from a standpoint of migrant and refugee rights, human rights, the right to asylum and the right to live. Therefore, migration scholars are worried about practices in these fields, and the tragedy of the thousands of people on the move who lose their lives on the borders because of smart border applications.

However, the picture is different in the field of digital identity. There are, of course, academics who call attention to the corporate agenda-setting aspect of the field and criticize the entire process. However, a reformist approach is more dominant. The importance of issuing ID cards to people, digitalizing these IDs and providing people with 'fast and efficient' access to services thanks to digital IDs draws more attention. There is, of course, a sizeable literature on how digital identities lead to class-, ethnic- and gender-based discrimination, and I will be reviewing them in due course. However, both outright defenders of digital identities and many others concerned about the discrimination they produce seem to agree that it is possible to come up with a technical solution that would be more secure, more crypto-based, less centralized, more user-controlled, self-sovereign and so on. Many technical teams in universities, start-ups, large corporations and NGO sectors are currently working to this end and testing and piloting their products (FATF/OECD, 2020).

Despite these mostly positive approaches (with or without reservations), when we evaluate the issue from the perspective of the manifestation and development of surveillance capitalism in border and migration management, analyse corporate agenda-setting and new hegemonic discourses and decipher the oligarchic structure described in previous chapters, we can see that work on digital identities offers rich and revealing examples. What is more, compared with fields such as smart borders, which are dominated by military/intelligence firms and state agencies and characterized by secrecy, the variety of projects and initiatives concerning digital identities is more interesting and relatively more transparent, fostering interdisciplinary discussions.

It is not possible to cover all aspects of this discussion in any detail, as it evolves and expands all the time. In this chapter, I will limit the discussion, in line with the overall aims of the book, to those aspects that concern surveillance capitalism and corporate agenda-setting. I suggest that smart border and digital identity applications do not contradict but complement one another. If the main issue in the use of surveillance technologies in border and migration management is data collection and extraction, smart border applications function, through data analysis, to decide who is allowed to cross the border and, after crossing the border, how far they are allowed to go. A digital identity, on the other hand, is where such data is stored, and thus very valuable for both states and private companies. Both are fields that accelerate the development of surveillance capitalism and where its products are tested. Moreover, they make it possible for an oligarchic structure to form, centred around tech companies and bringing together military/security agencies and financial firms or capital with the corresponding bureaucracies.

Identity and identification in the field of humanitarian aid

It would be best to deal first with the issue of identity and digitalization of identity. First of all, it is critical for everyone to be issued official identity documents (IDs) at birth. This makes it possible for the person to be recognized as a citizen (or resident) from birth onwards, enjoy basic rights and freedoms such as those concerning education and health and benefit from social opportunities and welfare. In the absence of an ID, if you are not registered by a government, you cannot claim rights and responsibilities in most cases, and face a very challenging and problematic bureaucratic process.

The UN reports that more than a billion people worldwide do not have official identity documents. This is included in the UN's Sustainable Development Goals (SDGs) and the UN invites states, businesses and civil-society organizations to take action in this regard. The number of undocumented people is a very large one, and from the perspective of corporate agenda-setting, they represent a big market. Documentation is not only about recognizing people's social, economic and political rights by states. It is also important for taking part in the capitalist process of production as a worker or supplier and being a consumer/customer. It is especially important for tech companies and tech-oriented industries (such as financial firms and phone operators). This is because it plays a critical role in meeting 'know your customer' requirements, which are intended to prevent crimes such as terrorism and money laundering and serve general surveillance purposes (Slavin, 2021). People who do not have official IDs or are unable to produce documents to prove their identity find it very difficult, if not impossible, to open bank accounts, buy SIM cards or access the internet.

There might be many reasons why more than a billion people do not have official IDs. Here, I will deal with two main reasons. The first is that states may be following a deliberate policy of not providing ID documents to certain minorities living on their territories, with the explicit aim of political exclusion. This is the reason why Kurds in Syria and Rohingyas in Myanmar are not able to obtain official ID documents. A second reasons is that states may not have the infrastructure, financial means or personnel to register everyone living on their territories. For example, in many African and South Asian countries, states are not able to access and register everyone due to factors such as difficult terrain and tribal and nomadic lifestyles. In the former case, it would be very difficult to overcome political discrimination due to sovereign rights, whereas, in the latter, it is possible to come up with solutions to increase state capacity. However, it is important not to overlook political aspects and factors related to underdevelopment and colonial legacies.

A related issue, one that is more relevant for the purposes of this book, is the documentation of forced migrants, asylum seekers and refugees. Even if they were issued official identity documents, people who flee wars or natural disasters often find it difficult to carry their IDs, title deeds or diplomas with them, or protect physical documentation throughout their arduous journeys. Even if they manage to do so, there is no guarantee that these documents would be recognized as genuine in the host country, which creates a slew of new problems. Moreover, it is possible to forge IDs, especially paper and plastic ones, or fraudulently present documents belonging to someone else. Or, one person in the family may have proper identity documents and the rest of the family may not. Delaying the registration of newborns is known to be a common practice, especially if they are girls, and living in rural and isolated areas makes registration even more difficult. There might be multiple people sharing the same name and surname. There might even be multiple claimants for the same title, each bearing genuine documents. For example, after ISIS was driven out of Mosul in Iraq, settling the disputes over house titles between people coming back to the city and current residents proved to be very difficult. In some cases, different people had Ottoman-era, Saddam-era or ISIS-era titles to the same house. Humanitarian aid organizations had to deal with issues such as multiple genuine titles for the same property, with each claimant being granted the right of use.

In short, people fleeing conflict zones or natural disasters, forced migrants and refugees first have to prove their identities in the host countries to be able to access aid, obtain work permits, attend schools or receive healthcare services. This is true both for people who have official IDs and for people who are unable to produce documents because they were never issued or were lost or destroyed. UN agencies such as UNHCR, WFP and IOM, in particular, are taking action on the ground to solve this problem. Of the countries that face large migration flows, those with a strong infrastructure, such as Turkey, are able to develop solutions in a rapid manner, issuing ID numbers to new arrivals, allowing them to enjoy certain social rights (such as education, healthcare and work permits), and granting access to higher education if people are able to produce valid diplomas. Many countries, on the other hand, are unable to provide identity documents to significant portions of their own citizens, let alone new arrivals.

Naturally, the first task awaiting organizations that provide humanitarian relief during emergencies is to identify aid recipients and, if people are undocumented, offer alternative forms of identification in order to register everyone. This is a prerequisite for distributing aid, setting up camps and providing certain social rights and opportunities. Coming up with fast and efficient solutions is critical in a world where tens of millions of people are in need of humanitarian aid, thousands of people face imminent death, hunger, thirst, illness and lack of security under conditions of humanitarian emergency, and physical infrastructure is either non-existent or heavily damaged in many countries.

Given the urgency and the scope of the problem, and given that mass migrations have multiple destinations as people seek better places to meet their shelter, security and other needs, the idea of digitalizing identity and utilizing digital technologies gains ground as it would allow tracking of human movements and provision of help and support to people regardless of where they may end up. In the alternative scenario, where only paper and plastic IDs are used, problems may include forged IDs, a single person using multiple IDs to obtain additional aid, and disadvantaged groups such as children, women and disabled people being forced to surrender their IDs. Digitalization helps to overcome these problems.

Besides registration and planning benefits, digitalization also facilitates cooperation and information-sharing between different organizations. The sharing of data and identity information by UN agencies, local or central governments and other humanitarian organizations can make the distribution and monitoring of aid more efficient and help to solve the problem of duplication, which refers to some recipients getting aid from multiple agencies while others do not receive any aid. People who leave Venezuela, for example, do not stay in border towns, choosing to go to different cities in neighbouring countries. Similarly, Latin American migrants to the US often cross multiple countries on their journey. In both cases, people on the move can receive aid in different cities and countries from organizations such as the Red Cross, and issuing digital IDs is proposed to register aid recipients, monitor the process and avoid unnecessary red tape.

In some cases, humanitarian aid organizations or UN agencies may avoid coordination for legitimate reasons (such as preventing harm by authorities or anti-immigrant groups), or coordination may be hindered by concerns related to privacy and consent issues or a desire to protect organizational interests (such as preferring not to work with a rival organization). As a result, people in Somalia, for example, use multiple SIM cards to receive the monetary aid sent to them on a monthly basis, and never need the card in question at other times because multiple organizations send monetary aid using mobile banking and every NGO provides a separate SIM to recipients. This also means that each organization ends up paying the mobile operators a hefty sum for duplicate services. Another example is seen in worrying developments such as the Taliban confiscating digital identities prepared by the US.

Digitalization enables agencies such as the UNHCR and WFP to provide regular aid to close to 100 million people. Similarly, Ukraine has an effective and reliable e-government system, which plays a critical role in providing shelter, jobs, education and healthcare to Ukrainian refugees seeking asylum in European countries following the invasion by Russia. People are able to download official documents proving their identity, diplomas, school transcripts and health information, among other things, and share them with authorities, which in most cases eliminates the need to investigate people's claims about their identities.

Digital identities in the economic and social integration of refugees

I have already argued that the main goal in using these technologies for border and migration management, within the framework of surveillance capitalism, is to reshape and reposition the global working class and to decide who will be allowed to migrate and how far and which sectors they will be employed in. One aspect of this is smart border applications. To recall some of the examples in Chapter 1, when an Afghan or Syrian person leaves their country and ends up in Turkey, they are positioned as cheap labour for the supply chains of global textile and food companies. A Nepali person allowed to go to the Gulf countries without a visa can find jobs in the construction and tourism industries. A woman from a rural village in the Philippines can migrate to Canada or Germany to work in domestic services. IT workers or healthcare workers from Turkey or India can migrate to the UK or US and obtain work permits. Young people from Pacific islands can migrate to Australia as rugby players.

Smart border applications and similar practices thus offer opportunities for predicting, identifying, monitoring and, at some point, preventing individual or mass migrations, be they legal or undocumented. The main criterion is that the new arrivals are positioned as cheap labour in the most appropriate localities for them to serve the needs of local and international capital.

However, smart border applications on their own would not be sufficient to turn migrants into labourers, pull them from the periphery to the core of the capitalist economy and make them a part of global supply chains. Digital identity applications play a role here in controlling what people do once they cross the border. Digital identity becomes important in monitoring people who (are allowed to) cross the border, letting companies or states examine their documents in making hiring decisions or issuing work permits, and facilitating social and economic integration.

Some of the main issues involved in social and economic integration are as follows: opening a bank account, buying a SIM card, accessing the internet, using online services to carry out obligations, becoming a consumer and customer, generating and extracting data on an ongoing basis, ensuring workforce productivity by providing access to healthcare services, and allowing adults in a family to work by enrolling children in school, which also serves to train future workers. It is in the interest of governments and companies for data to be stored on digital ID cards so that all these tasks can be identified, controlled and managed and various organizations and companies can examine the relevant data as needed.

Smart border and digital identity applications complement one another. This makes it possible for migrant people who cross borders with the expectation of joining the ranks of the working class (or become suppliers, start-up entrepreneurs, shop owners or students), become positioned as producers and consumers and generate surplus value and data. It also allows surveillance capitalism to monitor the entire process from undertaking the journey to settling in another country, integration into host society, participation in the creation of surplus value, and retention of ties to the origin country (in the form of remittances, visits and vacations, the arrival of new migrants, trade and so on).

Before moving on to the debates on digital identity in migration management and on monopolist companies that actively intervene in these fields and adopt surveillance capitalism as their business model, let me use an example to explain why digital ID applications are adopted for people forced to migrate under conditions of humanitarian emergency, the rationale behind these policies and how private tech and financial firms become key players.

Intervening in humanitarian emergencies with digital IDs

How to distribute humanitarian aid is an important question in a world where forced mass migration is a common occurrence because of wars, conflicts, poverty and natural disasters in many countries. People in need of aid number in the hundreds of millions, civil wars and conflicts seem to take forever to resolve, and after natural disasters such as earthquakes, it often takes many years to make the necessary investment in infrastructure and return to 'normal'. The most fundamental perspective regarding this issue, although it has become something of a cliché, is the humanitarian-development nexus, which refers to adopting a development-oriented approach as soon as possible after the emergency, and creating the conditions for people to stand on their own feet or to empower them (Letouzé et al, 2013; Palvia et al, 2018). Because sources of aid are limited and global politics, balance of power, and public attention dictate where aid efforts are concentrated, continuing to distribute aid for years is not a sustainable proposal. Thus, what is needed is to adopt an approach that aims to solve the local socioeconomic conditions in the countries where people take asylum, and create the conditions for migrants to live with human dignity. This is the medium- and long-term approach to be kept in mind.

However, there are many tasks that must be carried out in the short term as well, in the immediate aftermath of a humanitarian emergency. People should be able to stay in safe places, camps should be set up if necessary, and the conditions should be clarified if they are to live among the local population. It is only natural for vulnerable people with traumatic experiences – many of whom are children, elderly, disabled, or women – and who have suddenly lost their loved ones, homes and everything else that makes a life meaningful, to have many needs. Moreover, continued fear for one's life and an uncertain future make it more difficult for people to recover in psychological and sociocultural terms. Organizations and teams arriving on the field to distribute aid under these conditions must have a comprehensive plan in place. One main reason for that is to secure the supply of aid. Some localities may be remote and difficult to access; others may lack infrastructure or it may be damaged; there may be other dangers or threats on the route through which aid arrives; or local authorities may not be friendly.

Coming up with solutions to these problems is one thing but distributing in-kind aid can also have different repercussions for recipients. Problems related to securing the supply of packages for regular distribution to all can be solved one way or another, but distributing identical packages to thousands of people may not solve their problems. There is no need to provide, for example, sanitary pads to families without women, nor nappies to families without babies.

The usual solution to this problem is to provide cash aid. Distributing cash makes it easier to monitor and control aid. Because shops and markets would also be set up so that the money can be spent, people can make their own purchases based on their individual needs. This supports people's financial situation and offers opportunities to save up or use the money in ways that are most beneficial to them, even when the amounts involved are small. Moreover, when a market forms, the local economy is revitalized, various goods and services are prepared and sold, and forced migrants would start participating in economic activity as producers and consumers. For migrant people living among the local population, not in camps, cash aid would make it easier to participate in the socio-economic life of the locality, and, because they would be spending the aid money in the local economy, local shopkeepers also benefit.

For all these reasons, there is a case to be made that cash transfers are more advantageous compared to aid in kind. However, providing direct cash aid to thousands, tens of thousands, or in some cases millions of people in need of humanitarian aid is a risky and difficult undertaking. Carrying physical banknotes to these regions brings security risks. In many regions, gangs or warring parties may attempt to steal or seize the money. Secure transfer of cash requires a well-organized effort. Even if banknotes were distributed, people may find it difficult to spend them freely. Local authorities or security personnel may demand bribes, shopkeepers may raise prices, or gang-like organizations may seize the money. Moreover, women and young people, in particular, may not be able to spend the cash as they desire due to various factors such as sociocultural institutions, traditions and patriarchal/feudal/hierarchical relations, and may even reinforce existing inequalities and discriminatory practices.

In this case, if the option of aid in kind is not reconsidered, the digital transfer of money, via mobile phones, is a solution. This, in turn, solves other issues, such as making sure that people have phones and SIM cards with sufficient credit for SMS messages, access to the internet, can use mobile banking and have bank accounts. This is where phone operators, financial firms and tech companies enter the picture. Using mobile banking for transfers of money to millions of people on a regular basis requires having a comprehensive technological infrastructure in place. Solving this problem is beyond the capabilities of humanitarian aid organizations, the UN, and even the governments and banks in many countries. Digital transfer of funds from donors – in other words, from advanced capitalist countries – also makes it possible to track the money and solves, from the perspective of donors, the problem of auditing UN agencies and aid organizations by providing accountability and transparency.

On the other hand, in case of aid in kind to people fleeing war zones or natural disasters, identities of beneficiaries do not necessarily matter. You simply hand out a package to everyone arriving. However, if you opt for digital or mobile transfer of money, you first have to identify aid recipients. As mentioned at the beginning of the chapter, 'know your customer' requirements mean that a bank cannot simply transfer funds to any person and a phone operator cannot issue a SIM card to just anyone. They have to make sure that the funds are transferred and the card is issued to an actual person with a confirmed identity. Moreover, record-keeping is also important for donors to track where their money ends up and to predict the size and duration of support needed, so digitalization provides better opportunities on this topic.

This is where the debate around digital identities enters the picture. It is not enough for aid recipients to simply submit a document, a provider also has to make sure that they are actually the person they claim to be. This is a whole other issue that needs to be resolved, but, in any case, it is important to issue a new digital identity or digitalized identity information because this data will then be transferred to the phone operator and the bank so that they can send the money digitally on a regular basis.

This process requires a high level of organization. Internet access must be provided, mobile phone coverage should be extended, tools and training for mobile payments and identification made available, and the necessary decisions and investments made beforehand to distribute phones and SIM cards. This process creates the conditions for many private companies to enter the picture, especially finance and tech companies and phone/network operators.

Depending on the infrastructure available in the country where digital cash transfers are to be made, bank cards or debit cards can be used to undertake regular transfers of money, and people can use these cards for their purchases. Beneficiaries can receive codes by text message, and people can provide these codes along with their identity information to sellers to make purchases. Or, as in the Zataari camp in Jordan (Juskalian, 2018), eye scans can be used to access and charge refugees' bank accounts.

When digital transfers are used, it becomes very difficult for gangs to seize the money, for corrupt officials to take bribes, or for the vulnerable to be extorted. Digital funds offer better opportunities in this respect compared with physical banknotes.

Use of this method to distribute humanitarian aid and the rapid increase in the size of cash transfers as a proportion of all aid is encouraged both by donor countries and the big corporations involved in the process. Financial firms, tech companies and phone operators, in particular, are developing special lobbying groups and initiatives regarding refugees and forced migrants. In return for fast and efficient interventions, they provide, for many years, a wide range of critical services, from SIM cards to bank accounts to mobile banking to millions of people, using public funds.

This creates two sorts of customer for the companies involved. The first is the UN agencies and humanitarian aid organizations that choose mobile banking for cash transfers. They are offered the digital infrastructure for cash transfers. On the other hand, millions of people who receive the aid become customers of these companies as they use their phones, access their bank accounts and make purchases. This, in turn, means that the field of humanitarian aid, which has traditionally been characterized by public-spiritedness rather than profit-seeking, becomes commercialized and turns into a profit-making field. The company contracted to facilitate the aid meant for a certain group of people in a given country also gets the opportunity to enter a new market and access a new network of customers. Given how powerful these companies already are, it is highly likely that they would leverage the investment they make not only to provide services to aid recipients but also to establish ties with the local population and thus expand their market share.

To reiterate a point made in Chapter 1, aid organizations and UN agencies become data suppliers to global companies. Without employing a single person in war zones, conflict regions, areas hit by natural disasters, or places where people fleeing these calamities take refuge, these companies get to register thousands of people thanks to the humanitarian staff on the ground, who are also tasked with distributing the cards and collecting biometric data. They access new markets in a highly profitable and low-risk venture financed by public funds from donor countries and based on the labour of humanitarian aid workers.

Recording an entire life

I have offered a lengthy narrative on the problems experienced in the field of humanitarian aid and the digital solutions found, but I will not discuss the alternative just yet. I will do so in the Conclusion. As a result of the choices made so far, private companies have entered the picture and found new markets and customers. Capitalism has expanded its sphere of influence, making use of humanitarian aid organizations in the process. Let me leave this observation here and return to the debate about surveillance capitalism and digital identities.

First of all, these developments and decisions show that the issue of digital identities or digitalization of identity is not limited to identifying people or proving one's identity. Data stored in digital IDs are not limited to a few pieces of information on citizens collected by governments at birth (name, surname, date and place of birth, names of parents and so on). These IDs record the entire lives of the recipients of humanitarian aid. Thanks to the collection of data to identify people and enable them to purchase SIM cards, open bank accounts and make use of mobile banking, and given the demands of donor countries to audit the money spent in the name of transparency and accountability, the personal digital identity of a refugee contains data on all the aid provided so far. What is more, the ID cards in question also record all the expenses made. In addition, now that they have SIM cards in their phones, CDR data can also be tracked, which makes it possible for phone operators to access users' call history and make key observations about their socio-economic status and cultural characteristics. The digital identities store – or make it possible to record and monitor – all sorts of information about refugee people over many years, including their workplaces, the schools their children attend, and the certificates obtained by their family members, among others. In other words, the initial motivation of identifying people and thus facilitating their access to rights and responsibilities rapidly evolves into something else, becoming an infrastructure for surveilling and recording all moments of a person's life.

We have seen, for example, that the World Food Programme provides aid to 100 million people worldwide, and the digital infrastructure for that was provided by the infamous Palantir company of Silicon Valley. Keep in mind too that several financial firms and phone operators were also involved in this deal. All the data generated by these 100 million people is collected and analysed by these companies. This is clearly a gold mine for surveillance capitalism. These corporations make investments using public funds, and access new markets and millions of new customers. They earn more than the fees they are paid or the profits they make. They also gain access to data extraction, the holy grail for surveillance capitalism. And they do so in places where privacy-related laws are non-existent or not enforced and with minimal oversight, if any, which means a great opportunity for surveillance capitalism to move from being a mere business model to an intrinsic feature of capitalism. The firms gain access to data from millions of people that no other monopolist company has access to; people who have somehow remained outside or on the margins of capitalism, and the ability to examine their behaviours. As a result, different surveillance technologies continue to be tested.

For surveillance capitalism, digital identity represents a very productive and high-priority area of heavy investment, and for the monopolist companies that adopt the business model of surveillance capitalism, digital identity is more than a simple means of identifying people. Identifying people and issuing them ID cards is only the first step in the process, followed by years of data extraction.

Digitalization of identity and its relationship to surveillance capitalism

At this point, let me move away from the field of humanitarian aid and migration, and have a brief look at the overall picture and trends in the field of digital identities. This will allow me to show that the problem is not limited to one field, and that migrants and refugees are used as test objects for developing technologies – blockchain and biometrics, in particular – and products in the field of digital identity.

Digital identity concerns all of humanity, is indispensable for surveillance capitalism, and attracts much investment because it is considered to play an important role in the development of the digital economy. What I mean by 'identity' here is registration in order to make use of certain services and products, by providing unique information that enables unique and personal verification, such as a username and password, all of which are part of our digital identities.

Issuing an identity document is viewed as one of the main prerogatives of the state in modern capitalism, but today, private companies also issue IDs. For example, your bank records your information for online banking, and so does your university for your higher education, your employer for online activities in the workplace, Gmail or Hotmail for your e-mail account, and Facebook and Twitter for the use of their services, assigning usernames and passwords consisting of letters and numbers. Sometimes this is for accessing certain websites or viewing the articles published by a newspaper. Viewing the content offered by a publisher may be free, but you may still be required to register. As a result, people who have access to the internet have digital IDs issued by many organizations. We are required to store or somehow remember the usernames and passwords for all these domains. When such information is lost, stolen, or somehow acquired by malicious actors, we may suffer losses of varying magnitudes.

Because of the multiplication of the number of organizations issuing identities, a certain hierarchy also emerges among digital identities. For example, electronic government or government gateway and online banking accounts have the highest level of security, and we are expected to remember registry information for these accounts. Thanks to this security infrastructure, it is possible in some countries, such as Turkey, to use online banking to access e-government services. On the other hand, you may be able to use your Gmail account to access certain newspapers or websites instead of going through a separate and lengthy process of signing up. In this example, the Gmail account occupies a higher position in the hierarchy. Problems that would result if these identities were damaged, stolen or forgotten would be proportionate to their position in the hierarchy. For example, you may lose money if your online banking information is stolen. You may face extortion or threats if sensitive information kept by the state ends up in the hands of malicious actors. On the other hand, the consequences of forgetting sign-in information for social media accounts should not be as severe in most cases. There might, of course, be examples to the contrary as well. Malicious people may use your social media accounts to post messages that would put you in hot water or ruin your reputation, for revenge or extortion purposes, perhaps.

Let me stop giving cybersecurity advice here and stay on topic. In the case of humanitarian aid, migrants and refugees, digital IDs help in the provision of regular aid by registering people and helping them integrate into the host society, but this has already become a topic that concerns citizens of the host society as well. Common to both is the fact that personal information is requested and recorded to provide access to certain services, and then data collection continues non-stop via that person's devices. The data collected is analysed by the organizations, and the third parties that purchase the data, and used to categorize people (even if they avoid giving their real names to remain anonymous) and send personalized commercial and political messages to them. This is also a concise description of the business model of surveillance capitalism.

Identities, within the forms we are discussing in this book, can be divided into two types: foundational and functional. Foundational IDs are those issued by a government, proving who you are. Functional IDs are designed to access certain services or products, and are limited to designated goals. For example, a driving licence allows you to drive a vehicle on public roads legally. We obtain functional IDs to perform online banking transactions, check our social media accounts, send emails or read newspapers (Slavin, 2021).

There are a billion people without foundational IDs. Refugees and forcibly displaced people need foundational IDs to be registered and adapt to new conditions, but foundational IDs are insufficient for capitalism in both the field of humanitarian aid and other online spaces. The needs of surveillance capitalism and its demand for data analysis can only be met with functional IDs. The goal is to collect as much data as possible and utilize it for profit-making purposes.

However, there is a growing demand for privacy in many societies. Resistance is forming to non-stop data collection, categorization, predictive marketing and manipulation by companies and governments. Moreover, in the case of a humanitarian emergency, it is obvious that the people involved are in a vulnerable position. The 'no harm' principle is one of the most basic principles for UN agencies and humanitarian organizations. In other words, they must take measures to protect the people they help – who are often traumatized and fear for their lives – and to prevent any initiatives or activities that could harm them. Therefore, even though these agencies and organizations become data suppliers to tech companies (and finance firms and phone operators), in practice, they still demand that certain principles are followed and expectations met. This is the basis of the unique status of the debates around digital identity regarding humanitarian aid and refugees.

As we will shortly discuss with examples, the field of humanitarian aid and refugees is where the most effective work is undertaken on tech solutions that would meet multi-layered demands such as the security of digital IDs, data protection and sharing and privacy. A large number of NGOs, UN agencies, lobbying organizations, global corporations and start-ups are using the most advanced technologies to come up with the most progressive, radical and speculative positions in the debate on digital identities (Cheesman and Slavin, 2021). This is why biometric technologies and blockchain technology, which became well-known through association with cryptocurrencies such as Bitcoin, are common in the fields of humanitarian aid and refugees. These are work-in-progress technologies that have the potential to develop, but we have scant knowledge of their potential costs or disadvantages. They are again tested on refugees and migrants to get a better idea of their potential impact. The field of humanitarian aid and refugees is the site of a lively debate, utilizing concepts such as self-sovereign IDs and decentralized IDs, and where several start-ups and monopolist companies such as Microsoft develop and test products via pilots in the Global South.

Just as weapons manufacturers test their new surveillance products on migrants (as explained in Chapter 3), technology and finance firms test new and risky tech products such as blockchain and biometrics on refugees. For example, there is sizeable literature on the discrimination and problems caused by biometric technologies (Cheney-Lippold, 2011; Brinham, 2019). Many studies in this literature express their concerns by looking at the results of the pilot projects that carry out tests on migrants and refugees. Updated models are designed and implemented on the basis of these worries and criticisms. However, it is impossible to know whether the people subjected to a previous project were harmed in the process or not, and how they fared. Similarly, blockchain-based projects are tested on migrants and refugees with slogans such as 'user-controlled', 'self-sovereign' and 'decentralized', even though the beneficiaries are not able to grasp the working principles of this technology in any detail. In case of failure, they are replaced by a new model or a new concept. However, the exploitative neo-colonial attitude toward migrants and refugees remains.

I have argued that AI-supported lie detectors piloted on borders would spread to the rest of society if successful, being deployed in schools, workplaces and police stations; I believe border-security experiments conducted with sensors, robots and drones are preparation for future wars; and now, I assert the use of new technologies such as blockchain and biometrics by finance and tech firms to store and analyse data in 'suitable' conditions is piloted on migrants and refugees. Even when these products end up benefiting refugees, the entire process from initial conception to design, from budget allocation to product development and piloting takes place without any input from refugees or migrants; they are all conceived and implemented based on corporate agenda-setting and interests, for more surveillance and higher profits. Therefore, without going into the issue of success or failure and the development of better products, what I would like to underline at this point is that they are all products that advance surveillance capitalism and are supported by companies and organizations that are dominant actors in surveillance capitalism.

More cryptographic, safer and privacy-oriented products may result from this process. However, as long as the practice of non-stop collection of data and accumulation of data on all aspects of life remains, a safe product developed and marketed by a dominant actor of surveillance capitalism can increase the market share and profits of its developer, and while it may protect user data from simpler cyberattacks or surveillance by other organizations, it would in no way hinder the development of surveillance capitalism: on the contrary, it would deepen surveillance capitalism and speed up the process of monopolization.

I believe this issue can contribute to the debate on creating alternatives as well. Based on my personal observations at the University of Oxford, I can confidently note that many academic research groups are conducting valuable studies on issues such as disinformation, privacy and discrimination and developing related products. When they are tested at some point and found to be successful, the next stage is usually their commercialization. The goal is for the project to create social impact as an output. However, a contradiction arises when a study that reinforces privacy and uses cryptographic methods to strengthen data security is submitted to, say, Microsoft; a product that uses simulation to detect accounts that spread misinformation and fake news is given to Meta, or a methodology that aims to prevent online or AI discrimination is presented to Amazon. On the one hand, an important step is taken for the solution of a specific problem, but, on the other hand, when these products are handed to the very actors that create, develop and foster surveillance capitalism, they end up strengthening and deepening surveillance capitalism further and help monopolist companies to fortify their monopoly position and increase their profits. What is more, from the perspective of people who are worried about these issues, have high levels of awareness, join campaigns and force companies to take steps by publicizing these issues, it means expecting the solution from the very actors that have created the problem in the first place and keep supporting the status quo.

The business model of surveillance capitalism is essentially based on manipulation because it tries to predict the future based on data analysis, sells this future to clients for marketing purposes, and shows the products and services of those clients specifically to people who might be interested in them, bombarding them with an onslaught of messages and advertising. Given the general criteria of capitalism and the desire of surveillance capitalism to shape the future, and, given that this system creates unfair competition, accelerates the process of monopolization, has close ties to governments and military bureaucracies, and constantly reproduces discrimination, extraction of surplus value and inequalities, it is naive to expect the dominant actors of this system to prevent discrimination and disinformation and respect privacy.

Leading organizations

Let me emphasize my point that the issue of digital identity is on the agenda of developed and advanced capitalist countries such as the UK as well. Many companies operating in this field argue that it is indispensable for the development of the digital economy. The main reason is clear. According to a study conducted by Soprano (2022), 10 per cent of the population of the UK is unable to benefit from many online services and products because people are unable, for one reason or another, to prove their identities. This may be related to factors such as age, gender, ethnicity and class, and it is also possible that large numbers of people simply prefer not to deal with the hassle of identity verification. However, this is a market of 7 million consumers. To ensure their participation, products are being developed and marketed to facilitate the process, securely store identity information and allow people to share their information quickly at different times and in different places. When we look at the issue from the perspective of the role of digital identity in the development of the digital economy, it is obvious that the function of digital identity is not limited to identity verification, which is only the beginning point of a longer process. Once identity is verified, products developed under the name of identity are used to diversify and continue the data-collection process. This, in turn, reinforces the hunger of surveillance capitalism for more data and its efforts to extract ever more data, which are dominant characteristics of surveillance capitalism.

Within this framework, the World Bank is the most prominent organization that recommends and advocates for digital identity and undertakes work for the political and legislative infrastructure required, both in terms of the overall digital economy and more specifically in the field of humanitarian aid and refugees. As the champion of this process, the World Bank works with a missionary zeal to develop and spread digital identities. Many other organizations, such as the World Economic Forum, work in cooperation with the World Bank for digital identities and the digitalization of identity, and for the widespread and safe use of functional IDs. The EU also supports the digitalization of identity in member states, but the debate at EU level is also influenced by the needs and worries of national governments and authorities.

Beyond this, tech and finance companies such as Mastercard and Microsoft steer the debate on digital identities both through their own work and through lobbying and grants. The lobbying organization ID2020, established by several monopolist companies, including Mastercard and Microsoft, runs projects focusing on the role and importance of digital identity, particularly in less-developed and more-dependent countries and in the field of humanitarian aid and refugee movement.

Mobile phone operators constitute another lobbying group that cooperates with finance and tech firms. The GSMA, which brings together phone operators worldwide, runs many projects on digital IDs. Some of these project focus specifically on forced migrants and refugees. They work in close cooperation with UN agencies and humanitarian organizations, and with the participation of members in different countries.

I will demonstrate the point by noting important corporate actors that advance work on digital identity from the perspective of surveillance capitalism. It is possible to access many of the reports by these organizations using open sources. The actors that enter the picture and steer the process, develop and market products and cooperate with the UN agencies are the World Bank, ID2020, Mastercard and GSMA and its members including Vodafone and Orange. From the perspective of these companies, on the one hand, there are humanitarian organizations that demand the investments using donor funds and, on the other hand, millions of beneficiaries who have been receiving regular aid for many years and are offered different services.

The main document that forms the basis of this cooperation and provides legitimacy to the process is the Sustainable Development Goals of the UN. SDGs list the main problems faced by humanity, especially in the least developed countries, and recommend that governments, companies and civil societies work together to find solutions. Because issuing IDs to everyone and spreading digital IDs are included among these goals, the monopolist tech companies that operate at the global level are able to play an active and leading role in the fields of humanitarian aid and migration.

When it comes to biometric records, there is a competition between the EU and the US to have the largest database. This is not a rivalry, of course, because the sides visit one another, share their experiences and learn from one another. For example, eu-LISA and Frontex officials met with US Customs and Border Protection officials in Washington in June 2022, to learn about the pilot projects the latter ran on the US-Mexico border and examine its biometric database. The border management system developed by the Transport Security Agency using blockchain attracts much interest from European visitors. Similarly, on the US-Mexico border, the success rate for a system scanning people's faces while they stayed in their cars was 40 per cent, which increased to 76 per cent with the addition of a second camera (Monroy, 2022). Of course, Frontex had to replicate that success!

It should be kept in mind that all these agencies face heavy criticism. The desire to record everyone and create larger and larger databases brings the involvement of Big Tech companies, and they start marketing their products. Cooperation of this sort and the provision of products by a small number of monopolist companies provide important clues as to the modus operandi of the oligarchic structure that has been discussed.

The EU will have the largest police biometric database in the world with the European Entry/Exist system, due to go online in spring 2023; as if the existing Schengen Information System, Eurodac and Visa Information System were not enough. This database will keep four fingerprints and facial images of citizens of third countries. While the US keeps biometric data from about 275 million people, the new EU programme will aim to collect data from 400 million people (Van der Ploeg, 1999; Broeders, 2007; Monroy, 2022; Aus, 2006; Kuster and Tsianos, 2016; Prentzas, 2021).

The US continues to make new investments in this field too. The Department of Homeland Security (DHS) is taking steps to create a US$6.16 billion next-generation biometric database. This Homeland Advanced Recognition Technology System (HART) will process data from DHS, FBI, local police forces and other sources. Using military-grade technologies, this system will store data from facial recognition, DNA, iris scans, fingerprinting and voice prints, among others. The program is jointly developed by the weapons manufacturer Northrop Grumman and the private equity firm Veritas Capital. It will purchase data from companies such as Clearview AI, which worked for the Royal Canadian Mounted Police and was fined for developing a facial-recognition system using photos obtained from social media without consent. Coupled with social-media mining, the security agencies and firms involved will have access not only to biometric data but to data on many aspects of people's personal lives as well. This data, in turn, will be stored on Amazon Web Services GovCloud (Kelly, 2022; Mijente, 2022). This was supposedly developed for border security, but because other agencies have access to the system and it allows recognition of people in a crowd or in cars, there is nothing to prevent its use against people exercising their democratic rights, including the right to protest. For example, anti-insurgency tools based on the facial-recognition system developed in Iraq and Afghanistan were used against people participating in Black Lives Matter protests in Baltimore in the US (Mijente, 2019).

In the name of providing humanitarian aid and border externalization, the EU offers many countries infrastructure to collect biometric data. For example, the Emergency Trust Fund for Africa was used in 2016 to establish a fingerprint database in Mali and Senegal for €53 million. Civipol, the firm that set up the database, is a partnership between the French government (40 per cent) and the weapons manufacturers Airbus, Safran and Thales (Achiume, 2020).

Prominent technologies and their working principles

It is not surprising that the most advanced, original and creative debates regarding the digitalization of identity and the strengthening of the digital economy take place in the fields of migration and humanitarian aid because new technologies and products are first tested on migrants and asylum seekers. Therefore, the prominent approach in these debates is consistent with the overall picture described throughout the book. What is special about these people used for testing is that they are some of the most vulnerable and marginalized groups in any society. Therefore, a number of measures must be taken and worries addressed to ensure that they are not harmed. For example, if a centralized database is hacked, data stored in it can leak and lead to harm of migrant people. Or a centralized database can make it more difficult for different organizations to cooperate or share data because of legal obligations and red tape. This raises the question, 'Why not establish a decentralized system?', which leads us to the new technology of blockchain.

Biometric technologies can be a solution to the hassle of proving one's identity, the problem of lack of trust in documents submitted and the task of monitoring people's journeys. Thus, it would be sufficient for a person to have their eyes scanned or provide their fingerprints instead of submitting documents, and it would be impossible, at least in theory and in most cases in practice too, to fool the system.

As in the Zataari camp in Jordan, these technologies can be used together to check and make sure that the cash support provided is actually used by women themselves as the intended recipients of aid. To spend the money sent, refugee women first get their eyes scanned in the shop – that is to say, use biometrics to prove their identities – and then access their accounts using the blockchain wallets assigned to them and have their accounts charged.

If you do not want refugee data to be hacked or leaked, you can utilize blockchain technologies based on advanced cryptography. If you do not want to share data with banks or pay bank fees when providing cash support, you could use blockchain technology to perform cash transfers. Or, if you are trying to create a more democratic governance structure instead of relying on a hierarchical system of humanitarian aid and migration management, and thus achieve secure data sharing between different actors while resolving issues of transparency and accountability with regard to the donors, again you can consider utilizing federated governance opportunities made possible by blockchain technology.

In short, the idea of using these technologies has gained prominence for resolving many of the political problems and concerns in the field of humanitarian aid, migration and refugee management – the problems that have been in place for many years. In most cases, biometrics and blockchain can be used together. For example, first, you collect biometric data when registering people, taking their pictures, scanning their irises or getting their fingerprints. As a result, people are no longer required to prove their identities every time they apply for aid or shop, pulling up their biometric data instead. Then, they can access the blockchain wallets assigned to them and have their accounts charged. All these transactions are accumulated or stored in the digital wallet protected by the blockchain's cryptographic features. And, importantly, it is impossible to track these transactions by pre-designated organizations. The data stored, in turn, makes it possible to report to donor countries that spending is done efficiently.

Projects that develop and implement both technologies are actively supported by donor countries and agencies. So much so that this had become a hype before the pandemic, and there were examples where blockchain was used specifically to secure funding, although it was not mandatory. During an interview with a tech company, I learned of a case where the aid organization was unable to access the information protected by blockchain encryption, and asked the company providing the technical infrastructure to remove the encryption and provide the data in a simple spreadsheet format. This renders security and privacy safeguards meaningless, even when they are used as a justification for using the technologies, not to mention all the possible dangers and downsides of blockchain technologies, such as how inefficient it is in handling multiple transfers or how high the energy cost is.

Ultimately, what matters is the donors' attitudes. Leading donor countries such as the US, UK, Norway, Sweden, Germany and Japan require innovation in many project applications. The requirement to develop innovative solutions for fast and efficient distribution of aid, corresponding to keywords such as security, privacy, empowerment and integration, forces UN agencies and aid organizations to include technologies such as blockchain, data analysis and AI as they prepare project proposals. This has two main consequences. First, because of the limited time available for applications and competition between organizations, the desire to eliminate rivals using these buzzwords and promises gains prominence, and projects are designed and submitted in the headquarters of aid organizations. This means that projects are not created in a bottom-up process, based on an analysis of concrete facts, which would require understanding problems on the ground and coming up with genuine solutions, involving migrant and refugee people in the process, perceiving their real problems, and taking action on their initiative, or, at the very least, understanding the expectations of the local representatives and employees of these organizations. Second, because aid organizations are not tech companies and donor funding is usually available only for short-term projects, they cannot turn into organizations that invest in and produce technology. Therefore, they have no choice but to cooperate with tech companies. Some of these may be small- or medium-sized firms. However, mobile operators, financial firms and large tech corporations also enter the picture – the bigger the partner, the better the chance of receiving funding for the project. A brief online search produces long lists of projects in the fields of humanitarian aid, migrants and refugees that involve phone operators such as Vodafone, financial firms such as Mastercard and tech companies such as Microsoft. In the case of tech firms such as Microsoft and Palantir, it is also possible to find joint initiatives they undertake with the border and migration management agencies and security bureaucracies of various states. Together, these actors cover different pieces of a puzzle which is made sense of when focusing on surveillance capitalism.

In this sense, the oligarchical structure described in Chapter 3 as being a product of the close cooperation between military/security/intelligence agencies and technology and weapons companies is complemented by the close cooperation between finance and tech firms on the one hand, and the development, foreign policy and economy bureaucracies of donor countries (which tend to be developed, advanced capitalist countries) on the other. This, in turn, prevents a significant proportion of the humanitarian organizations – which rely on donors' funds for their operational income – from producing alternative policies and projects that would shake up the status quo. In any case, such firms play a dominant role in 'innovative projects', from conception to design and implementation. In practice, this goes well beyond being partners in a project. Thanks to their technological superiority, the companies involved end up heavily influencing the project from the design of technological products to their implementation, in order to keep the promises made to donors.

The role of the humanitarian aid body, on the other hand, is limited to explaining the goals and the situation on the ground and training its employees so that the product can be implemented successfully, meaning that they end up in a secondary position despite shouldering the heavier burden. Therefore, blockchain and digital ID projects are seen in many different regions of the world, in countries such as Nigeria, Kenya, Nepal, Turkey and Thailand, in wildly different conditions. In some cases, efforts are made to create the infrastructure for blockchain in places that lack even rudimentary internet access or appropriate expertise in the relevant technologies.

As digital ID projects are designed to prioritize corporate interests, many of them fail to be implemented in line with the initial goals stated. For example, for blockchain technology to be decentralized and have a federated governance structure, people and organizations involved must have public and private keys. These keys are passwords that make cryptography possible. People and organizations use their public keys to send data, funds, contracts and other documents to one another. In this sense, the public key is like a public address that we share with others. When we receive a message or transfer, on the other hand, we can access the message in question using our private keys. Because private keys are kept by individuals, no one other than the intended recipient would be able to access the message. The transparency of blockchain refers to the fact that all transactions on the same chain can be viewed in the form of blocks, and it is visible who is sending messages to whom. Thus, the combination of public and private keys enables identification, and the verification problem is overcome because all messages are attached to one another in a chain. Security and privacy, on the other hand, are ensured by the fact, thanks to encryption, that messages can only be viewed using the private key.

Even though this technology was developed and popularized via cryptocurrencies such as Bitcoin, the potential of blockchain to be used in diverse fields is a frequent topic of discussion. This can be tested in the field of humanitarian aid and refugees. To use the Bitcoin example for added clarity, if you are paying online platforms to make transactions, these organizations do not share private keys with you. And when you are not given the private keys, you do not have real ownership of these Bitcoins. You are paying these companies to make purchases on your behalf, assuming they will sell the assets in question and transfer the proceeds to you when you tell them to. You hope that they will hold up their end of the deal. If you want to buy or sell during rapid market fluctuations, there is a chance that you may not be allowed to do so because you do not possess the private keys. As a result, there have been many examples of operators of these digital platforms suddenly shutting up shop and disappearing. News of corruption and scams are not surprising to people who know how the process works because the platforms holding the private keys have de facto possession of the bitcoins purchased on your behalf.

Therefore, what I would like to emphasize, without going into the debate on the regulation of cryptocurrencies, is that all parties involved must have a certain level of digital literacy and a good enough grasp of blockchain technology, in addition to the technological infrastructure required, for blockchain to be put into active and beneficial use. This is the only way for the transactions to be visible to everyone but the contents only to the holders of private keys when parties transfer money, contracts, or documents to one another. This would make it possible to conduct business without going through a centralized and hierarchical structure, a middleman to provide trust (a state, central bank, notary and so on) and having to pay fees or taxes in return for their services. Public blockchain networks such as Bitcoin and Ethereum could also be used – for example, Etherium was used in a humanitarian aid project after the Nepal earthquake – but it would be a slower process. Or, a closed system that is faster and more confidential could be established, with only the relevant people and organizations registered. The WFP project in the Zataari camp is an example of a closed system.

In short, the decentralized governance model offered by blockchain needs certain conditions to be met, and not all of these are technical. In the field of humanitarian emergencies/assistance and migration management, relevant organizations and the individuals designated as beneficiaries must have their private keys and perform transactions using their keys. However, under emergency conditions, in refugee camps or even under normal conditions outside camps, it is very difficult for people to store long passwords securely and use them as necessary to receive the funds or documents sent. Many people may lack the required literacy or the conditions needed. Moreover, because every blockchain transaction is recorded in the chain and simultaneously becomes visible to everyone in the system, it is not possible to correct typos or other small mistakes by undoing actions. Attempting to do so may cause confusion because the erroneous and cancelled transaction would also be included in the blockchain. Another important issue is that, once the private key is lost, it is impossible to access those transactions. The media reports cases of people who are euro millionaires as they own lots of bitcoins but are unable to access the money because they have forgotten the private key, spending all of their allowed attempts except the last one for fear the money will be gone once and for all, and living with the hope that one day they will somehow remember their private key. One solution to this problem is delegation, that is to say, a trusted organization storing everyone's private keys, making it possible to recover them if lost, but this goes against the logic of the process.

How does this work in practice in the field of humanitarian aid? What was clear from my observations of projects run in two African countries, and interviews I conducted with a tech company that offers the technical infrastructure to humanitarian organizations in many parts of the world, is that private keys are not shared with the parties or the beneficiaries. In some cases, all the private keys are kept by the tech company, which then performs actual transactions individually on behalf of all stakeholders. The problems involved are multi-layered. To begin with, it is not reasonable to provide private keys to thousands or tens of thousands of people and expect them to carry out transactions without a hitch. A significant risk is that correcting small mistakes is not possible. Therefore, the target beneficiaries of the project are not able to access the digital identities created using the blockchain. In many countries, local banks, phone operators and humanitarian organizations do not have sufficient expertise regarding this technology, and their private keys are also kept centrally – by the tech company involved in the projects I have observed. What is more, the – often Western – humanitarian aid organization that is the main owner of the project may also lack personnel with sufficient knowledge of this technology.

In short, a project that receives funding with the promise of being decentralized and transferring aid to people in need in a timely manner ends up being the most centralized digital aid distribution system ever. If a central database not using blockchain were to be used, concerns such as hacking would still be present, but, at the very least, several organizations would be able to participate in the process using their own resources. Once blockchain is involved, this opportunity is also lost. There are so many potential problems, it would be difficult to create an exhaustive list. For example, when exclusive authority is given to the tech company, which is then expected to undertake all of the practical work, it creates a heavy and unnecessary burden for the company, and the humanitarian organization may end up handing out beneficiary data to an unknown company.

What is even more concerning for our purposes is that the digital identities given to large numbers of refugees and asylum seekers do not, in fact, belong to them. They cannot view or check the data accumulated on their digital IDs. This proves, at the very least, that migrant people's desires and preferences are not respected at all even as they serve as test subjects for these technologies.

Let me open a parenthesis here. The projects I am talking about are independent, funded by donors and run by UN agencies or humanitarian organizations in various countries. The issue of digital IDs based on blockchain technology is on the agenda of the EU as well, and a department is undertaking work on this topic. Because of the involvement of governments and public agencies, private keys might not be shared; or, even if they were shared, the EU and member states can choose to store the keys themselves based on their sovereign rights. In another project that I was involved in, Istanbul Metropolitan Municipality (İBB) started work on blockchain-based digital IDs for all residents in 2021. Central and local governments and public agencies, such as central banks, may put the issue of blockchain-based digital IDs on their agenda with different motivations. Administrative goals, such as secure storage of data and allowing individuals to access, check or edit their own data, may be more prominent in these efforts. Or, a centralized and distributed governance model led by a central bank may be preferred over a decentralized and distributed system. This may be proposed as a more efficient, transparent and democratic model of management.

What matters, however, are the presence of public authorities, the expertise of the participating agencies regarding the relevant technologies, and the details of the legal framework that applies. Our research project, which examined the digital identity initiative of the İBB, looked at how refugee people who gained access to these digital identities – which stored data collected by the municipality – could use the data for financial integration. Although there were many problems, the ability of people to check their personal data collected by public authorities and the presence of a clear legal framework on issues such as privacy were among the positive aspects of the project. However, these issues are experienced at a whole other level in projects implemented among forced migrants in Global South countries. The lack of rules and prevailing uncertainty in these places turn them into ideal testing grounds from the perspective of capitalism.

Similar problems can be encountered regarding transparency and accountability. When beneficiaries and implementers of a project do not have the required technological literacy, they find it more difficult to detect potential problems. Moreover, when a closed blockchain system is used, independent verification of the level of success and manner of implementation of the project becomes impossible, and problems remain invisible. Accountability to donors may be improved, but the project remains invisible to the public.

Creating separate wallets for men and women for cash transfers via biometric records and blockchain in some projects, while not allowing families to have joint accounts is problematic and reinforces gender inequality. A bottom-up approach that utilizes persuasion, education and organization should be employed to oppose discrimination and empower marginalized groups, but, when a colonial, orientalist approach is adopted to make unilateral decisions and force them upon people via technology, in a way that evokes the notion of 'white man's burden', it could create problems. For example, spouses in many families may want to have a joint spending or saving plan for the limited aid or income they receive. In many families, women may be responsible for managing the household budget. Unilateral approaches ignore differences on the ground.

In short, blockchain technology offers important opportunities regarding certain political preferences, governance structure, the issue of reform or rejection of central authorities, and developing peer-to-peer, decentralized approaches. However, it is very difficult to obtain outcomes that match these promises in the fields of humanitarian aid, refugee situations and migration management. In many of the largest and most prominent examples, the result was the opposite: more centralization, less transparency and reduced accountability.

Data minimization, data security, consent, privacy and alternatives

In each chapter of this book, the discussion leads, one way or another, back to the issues of migrant people's consent, privacy and data security. In this respect, digital identity shares the same basic problems that all data analysis and smart borders. Non-stop collection of data from migrants and refugees and using them as test subjects for new technologies against their will and limiting them to the roles of employees and customers offer many opportunities to surveillance capitalism. Capitalist actors get to access new sources of data as well as new markets and customers, test their latest products, and use the results to improve their surveillance of entire societies and reinforce their monopolist position. As in the field of smart borders, pilot projects in digital identity are financed through public funds provided by donors, implemented in many regions of the world thanks to aid organizations, and help to develop new tech products for more intrusive surveillance.

This concerns people working in the fields of humanitarian aid and migration. Controversies regarding the issue of digital IDs, in particular, are reportedly leading to polarization in the field. An article published in 2021 in the journal *Big Data & Society* by leading experts and academics in the field (Weitzberg et al, 2021) notes that this polarization does not generate healthy dialogue, and proposes to have a healthier discussion to overcome problems as digital identities give rise to certain worries in addition to their many benefits. This is a typical example of the reformist approach mentioned at the beginning of this book. Similarly, researchers from many universities are conducting work on 'trustworthy digital IDs' in the Alan Turing Institute of the UK. Many teams are working on developing more cryptographic, more secure and more user-friendly products. Concepts such as self-sovereign identity feature prominently in the discussions that accompany this technical work. What is meant by self-sovereignty here is that a person owns their own identity using a decentralized methodology and deploys it as they wish. The argument is that this would allow them to control their data, and data sharing would naturally involve consent. The potential success of this argument depends on the recognition of self-sovereign IDs by governments and private companies. Moreover, generating and using these self-sovereign IDs require a certain level of technological literacy and socio-economic status. For people who lack access to phones or the internet, or who are subjected to gender/class/ethnicity-based discrimination, having access to and making active use of these IDs would be very difficult.

The search for alternatives leads to the emergence of new concepts, which in turn form the basis of new technical work. However, as Cheesman and Slavin (2021) observe, many start-ups founded with this initial goal – trying to establish a system that would prioritize refugee people and allow them to control their own data – have failed, while some of them managed to remain in business by contradicting their initial principles.

The polarization and failure of these start-ups, and the disappointments faced in many digital-ID-oriented projects, mentioned earlier with examples, stem from a discussion with an exclusive focus on technical solutions. Because the problem is not really a technical one, hoping that it would be solved if only we were to use more advanced techniques leads to new mistakes. Or, coming up with a more advanced and more successful technical solution leads, in the process of its commercialization and implementation, to the deepening of surveillance rather than improved privacy. Examples include the technological products turned over to monopolist companies such as Microsoft and Amazon, noted earlier.

Therefore, I believe it would be useful to approach the issue from the perspective of surveillance capitalism and the overall trajectory of capitalism. We need to discuss how digital identity, even though it solves some of the problems faced by migrant people, is designed, developed and put on the agenda in line with interests of tech companies that adopt the business model of surveillance capitalism. Public approaches, like those of the EU and İBB examples, may offer hope in terms of data security, accessibility and transparency, but the root of the problem is the expectation that short-term projects which humanitarian aid bodies undertake in partnership with tech companies, without public oversight, would be a panacea. For example, engineers and decision-makers in the military/security/intelligence sector probably avoid describing themselves as merchants of death, and think their efforts are meant to protect against external threats and reinforce national security. This is why they are trying to detect migration movements using a range of methods, from satellites to drones and from sensors to social media analysis. In financial integration, distribution of aid and related fields, the process of digitalization of identity is associated, in the name of being 'pro-migrant', with positive goals such as development, access to basic rights, empowerment, gender equality and prevention of exploitation, and many academics, engineers and experts try to develop more advanced technologies and new concepts to solve these problems.

Leaving intentions aside, the issue is not defined in terms of the interests of migrants or the local society in either field. It is the big companies that initiate and steer these debates and conduct lobbying activities. Thanks to support from the relevant bureaucracies of donor countries, corporate agenda-setting acquires the power to influence civil-society organizations and academics working in this field as well. The common denominator of all work conducted in these fields is the development and testing of surveillance technologies, using migrant people as test subjects.

Privacy and consent no longer apply to migrants and refugees in the fields covered in this book. Actors responsible for that are the tech companies, finance firms, weapons/security companies and phone operators that adopt the business model of surveillance capitalism and dominate the process using their political and economic might. We are talking about a field that contains hundreds of millions of people including legal or undocumented migrants, travellers with tourist or business visas, forced migrants, asylum seekers and refugees. If we were to include people in poverty and those who lack official identities, the count would be in the billions. This huge 'market' offers the conditions for surveillance capitalism to shape the overall trajectory of capitalism. Surveillance has become a very comfortable playground for capitalists for many reasons, including the relatively limited recourse that migrants and refugees have to legal resources, insufficient regulation and implementation in Global South countries, the strengthening of the borders of the Global North, and the strong negative reaction to immigrants in many societies. All of these make it possible to support the most controversial projects for in-depth surveillance, conduct tests and develop new products.

In the case of biometric and blockchain technologies, which are the most prominent in the field of digital IDs, refugees or people in need of humanitarian aid cannot voice any objections or avoid giving consent, given processes such as formulating these projects, obtaining funds from donors, developing the tech products, having the infrastructure in place, training the personnel who will implement the project on the ground, and transporting the required equipment. Consent requires that people who refuse have meaningful alternatives. In this case, the only alternative available to those who would not like to share their biometric data is to refrain from receiving aid and avoid applying to humanitarian aid organizations. This, in turn, exacerbates the problems and the risks they face.

On the other hand, it is not just migrants but many people in general who find it difficult to understand these technologies, evaluate their pros and cons and make decisions accordingly. In many projects, officials briefly explain the aims and the methods and technologies used, but we cannot hope that this will be enough to produce meaningful consent. It is even debatable whether decision-makers and managers of the humanitarian organizations running these projects have a good grasp of these technologies and base their decisions on sufficient knowledge. Factors such as the need to secure donor funds also make it difficult for many of these organizations to undertake alternative work that would undermine the status quo.

Moreover, from the perspective of tech companies, issues such as how new technologies like blockchain will evolve in the future, what new problems will emerge, and how long the level of cryptography that is available today will remain in place are unknowns. Thus, it is clear that many projects that use the most advanced technologies are shaped by donor countries, big corporations and organizations such as the World Bank, but refugees, aid organizations, donor countries and even tech companies lack sufficient knowledge of their potential consequences and downsides. In the absence of such knowledge, it is impossible to make informed decisions about them, which is why they are being tested for potential downsides.

The same thing applies to privacy. These projects constitute the biggest evidence that privacy no longer applies in the case of migrants, refugees and recipients of humanitarian aid. Blockchain and biometric technologies do offer advantages in terms of the collection, storage and sharing of data, but they usually address organizational or corporate concerns rather than having a migrant orientation. Practices such as turning over private keys to private companies and failure to exercise sufficient oversight of companies show that the real focus is on seeing whether the technologies are used efficiently and detecting their potential weaknesses. The security of refugee data is, at best, a secondary concern. There are of course NGOs, private companies, humanitarian organizations and agencies that view this as a problem and try to come up with solutions, but these efforts do not have the power or the momentum to turn the tide. They are not in any position to challenge corporations such as Mastercard, Microsoft and Vodafone or organizations like the World Bank and the World Economic Forum. They can make their voices heard only in academic and civil-society circles or through peer-reviewed journals and trade magazines.

I believe debates about surveillance capitalism and the overall trajectory of capitalism are helpful in analysing issues related to digital identity. An alternative might be to work with public agencies instead of just using the latest technologies, and avoiding, as a matter of principle, projects that serve the aims of corporate agenda-setting, especially those promoted by companies that adopt surveillance capitalism as their business model. It might be useful to couple this with the relatively recent idea of data minimization. Various other principles can also be developed, such as collecting as little data as possible, collecting data only when needed and respecting the right to access data.

Conclusion: How Can We Resist?

I want to use this opportunity to briefly summarize our discussion until now, and provide some lasting examples, to contribute to my overall argument. In this book, I have examined the use of surveillance technologies in migration, border management and humanitarian aid, with the aim of explaining contemporary capitalism's mainstream approach to migration at a global level and the motivations behind the investments of tech companies in these fields. What is critical here is that the development and implementation of surveillance technologies in migration and border management have repercussions that extend far beyond these fields. Given the far-reaching effects of these technologies – encompassing lie detection and military robotics, and the prediction and prevention of mass movements – it is evident that they will soon have consequences for humanity as a whole.

In gaining an understanding of this potential, we also recognize that this topic is directly related to the evolution and future of global capitalism, and that surveillance capitalism is not simply a business model. The capabilities provided by these surveillance technologies and their use for mass surveillance have become an intrinsic component of contemporary capitalism and are actively tested on people on the move.

This process is led by an oligarchic structure based on close cooperation between the security, intelligence and financial bureaucracies of a small number of advanced capitalist countries and a small number of tech producers, military organizations and financial companies. Donor countries provide funding in the name of humanitarian aid, and, through academic projects or research and development, new and usually speculative surveillance technologies are developed using this public money. These are then tested on migrants and refugees in different countries by security bureaucracies or various organizations such as UN agencies, and products found to be successful and effective are commercialized and marketed to different states. The reason I refer to this as an oligarchic structure is that there is a network of people operating on different platforms in a small number of countries, including the US, EU, UK and Israel, who work together, share their experiences and frequently move among companies, and the decisions made by this network are implemented at a global level.

What makes the fields of migration, border control and humanitarian aid stand out is that they are the lowest-hanging fruit. Migrants, asylum seekers and refugees have a legal status that is much weaker than those of citizens, and states claim the authority to control those who cross their borders by invoking their sovereign rights. Moreover, the fields of border and national security are more secretive, face less scrutiny and leak less information, and, in today's world, societies take a less compromising and sympathetic approach to migrants and foreigners. This allows tech firms, along with military and financial companies and security agencies, to do as they please in the international waters of the Mediterranean, refugee camps and regions suffering from wars, famine and natural disasters, and implement the most speculative AI or blockchain projects.

An analysis of this structure, its connections and the actors involved reveals important clues regarding the approach to be taken and the alternatives. First of all, the issue is not limited to migration, border management or humanitarian aid. Examining the pilot projects being conducted in these fields makes it possible to contribute to debates in a number of different disciplines. The examples discussed in this book can be tied into studies on the future of work, warfare, artificial general intelligence, surveillance and democracy. For example, blockchain projects run in different regions to support refugee people and strengthen their financial integration, as well as solutions developed in response to concerns such as privacy, are likely to play an important role in the near future for finance and cryptocurrency industries, as well as for the central banks and public authorities that have an eye on this technology and the opportunities it can bring to governance of the financial system.

Similarly, robot dogs capturing migrants on borders, cameras and sensors that can detect movement from miles away and distinguish between humans and animals, satellites and drones that can monitor all movement on Earth, algorithms that can detect and predict mass movements, social media and smartphone analysis tools and facial-recognition systems and emotional AI developed in 'high-tech' migrant camps are all harbingers of how wars will be fought in the near future, and indicate how societies can be controlled and manipulated by authoritarian governments. I believe, therefore, there must be better dialogue between those involved in migration and those involved in a wide range of disciplines, while debates in other fields should be informed by analyses of the new technological surveillance products being tested on migrants and refugees.

Another related issue is the attitude we should adopt when discussing companies, academics and organizations that use migrant and refugee people as test subjects when running these projects. There have been several articles and reports that level criticism at these speculative projects, and almost all are united in their opposition to the testing of new technologies on migrants and refugees against their will, given the unknown consequences and potential effects. It is only right to criticize companies that develop and test their speculative ideas in this way and which, to this end, seek funds and support from donor states, the border security bureaucracies of various countries, EU and UN agencies and aid organizations. It is also right to challenge academics who take part in research and development projects for such technologies as lie detectors and to be involved personally in their development, as well as universities that – when encountering project proposals that would struggle to pass a strict ethical review and most likely be rejected by social science disciplines – seek to take advantage of a more limited and technical-oriented ethical approval process through other disciplines, as well as the academic institutions that fund such projects.

We should, however, not be content with limiting our criticisms to these visible institutions, and the companies and organizations exposed as acting in bad faith, as there is a systematic approach and alliance involved. There are, of course, differences in the effects and consequences of those who develop lie detectors and military robots and those who aim to improve financial integration via blockchain, although they adopt the same systematic approach and essentially serve the same purposes – exploring more effective surveillance, the manipulation and steering of societies and higher profits. It is, therefore, not a paradox when Palantir or Microsoft work with military corporations on smart borders while offering data analysis and cryptographic digital ID products to humanitarian organizations. From a capitalist perspective, these are complementary fields that help decide whom to allow to migrate, up to what point, and where surveillance technologies are implemented (Lyon, 2009).

Limiting ourselves to exposing a few companies would mean overlooking the role played by humanitarian organizations, UN agencies and the security and intelligence bureaucracies of the donor countries (advanced capitalist countries) that approve the products, or who approach the companies for support for their own projects. In short, we are talking about a process in which ideas are brought to life and implemented in different regions involving all 'stakeholders' in the sector who provide financial or practical support on the ground. It is, therefore, not enough simply to point out that a product is being tested, or to criticize a company conducting the tests or running the projects. When we associate this with capitalism, the role of migration in the overall direction of the system becomes all too clear.

The security-oriented approach to these fields has long been a matter of debate in academic circles, and the securitization of migrants and refugees in particular has been criticized. For instance, there is a large body of literature on the 'securitization of migration management' and 'externalization of borders' (Andersson, 2018). In this regard, much research has been conducted into the relationships between UN agencies, humanitarian organizations and states, as well as the hierarchical structure in which the advanced capitalist countries are dominant, along with the transparency of the projects and the effectiveness of the money spent in terms of actually solving problems (Cheesman and Slavin, 2021). What I would like to emphasize here is that the effects of the global-level system in migration, border management and humanitarian aid are not limited to these fields. These projects have consequences for all of humanity, a point repeatedly made throughout this book. Beyond their immediate effects and consequences, the projects, policies and investments in these fields have consequences that are directly related to the overall trajectory of capitalism and that complement one another. Because they are versatile and interdisciplinary, it is easy for technological experiments in these fields to spread to other fields and be rapidly commercialized.

It would also not suffice merely to point out the relationship between border and migration management and humanitarian aid, on the one hand, and other fields on the other, nor just to clarify their places in the overall structure of the capitalist system. It is true that companies that test and develop products in these fields make great profits from this activity itself, but when they do so with public funding and academic support, they can earn huge profits without taking any financial risks. Military companies that produce the military surveillance technologies tested in Palestine, the Mediterranean or on the Mexican-US border naturally enjoy advantages over their competitors in the market (Talbot, 2021; Tazzioli, 2021). Companies that facilitate cash transfers to millions of migrants, that issue cryptographic digital identities to them and that provide the infrastructure for data analysis also benefit in terms of maximizing their profits, reinforcing their monopolist positions and gaining access to new markets. If we were to evaluate capitalist policies related to migrants from the perspective of the global working class, it is clear that these companies are not the only beneficiaries of these policies and practices, as capitalist companies operating in many other industries also benefit thanks to the international division of labour. As discussed in Chapter 1, fast-fashion brands take advantage of the large migrant flows in Turkey and Jordan; large construction companies utilize migrants from South Asia in the Gulf nations, and migrants from central Asia in Russia; while Afghan people, not desired in Europe, are expected to work in agriculture in Turkey. It is possible to migrate from Turkey to Europe, or from India to the US, if you are a software developer or a medical doctor.

This brings us back to migration and border technologies. It is surveillance technologies that make it possible to allow mobility when desired, in a fast, efficient and optimum manner, and to identify, categorize and analyse new arrivals before they even arrive at the border. This applies to regular, legal migration and irregular migration. In the case of irregular border crossings and asylum applications, there are technological products available for cross-border monitoring that can also be utilized as war technologies. In the case of those who would like to arrive via legal means, by obtaining work or residence permits or visas, there are AI algorithms to analyse every bit of information that they offer about themselves, as well as that garnered by biometric technologies and lie detectors deployed at border crossings. These dictate whether they can move and how far.

Once we make that conclusion, we can achieve a more in-depth understanding of the approach to be taken in migration policies. Researchers and NGOs frequently discuss and criticize the hostile attitude of states to migrants. Former UK Prime Minister and Home Secretary Theresa May talked openly about creating a 'hostile environment' for migrants, while global talent, start-up, scale-up and entrepreneur visa policies made it clear that a warm welcome awaited those with the desired skills in certain industries. Once a person obtains endorsement certification from recognized and regarded institutions in the UK, they are considered highly qualified and will be given a work permit, without a sponsor, followed by permanent resident status in three years and citizenship in four years. Some countries offer citizenship when you purchase properties, while others are quick to provide work permits to temporary and seasonal workers. In short, in advanced capitalist countries in particular, migration policies are shaped in line with the needs of the authorities' view of the labour market. Legal steps are introduced to facilitate these policies, and no lengthy interviews or procedures are involved as, thanks to data-analysis technologies, the process can be run much more efficiently. Coordination among the relevant bureaucratic units and companies makes it possible to rapidly identify people with the desired skills and to speed through the necessary migration processes.

It is not sufficient, or even possible, to make a distinction between good tech and bad tech when examining capitalist policies regarding migration, or, in other words, the shaping and positioning of the global working class. Torture inflicted by security forces is of course unacceptable, as are decisions that lead migrant people to their deaths. However, migrants that are allowed to cross borders, that are given the right to work, study or reside in a new country, or that are invited and given other opportunities due to their particular talents, are also surveilled, and this process remains in full swing, from biometrics to data analysis and to the invasion of privacy via the required documents. Those who manage to cross the border may find it difficult to access their rights or complain about wages. In the strikes by academics in the UK in November 2022, university administrations warned those who were migrants, working on temporary work permits or who needed the university's sponsorship for their work permits not to join the strikes. Their warnings were, naturally, very effective. When you are an academic, software developer or physician from a developing country, landing a job or passing exams in the UK, US or Germany may be considered a great personal achievement, for you, your family and your community, but once you migrate and join the workforce, you face low wages and little opportunity for negotiation over your position, as well as non-stop surveillance.

As discussed in Chapters 3 and 4, it would be naive to talk about good and bad smart border applications, biometrics or blockchain technologies. Different means are used by the same entity – the oligarchic structure – to develop surveillance capitalism. These are tested and evaluated with the same criteria in mind, and the same technological product that facilitates faster border crossings in Norway may lead migrant people to drown in the Aegean. There is obviously a tragic difference in terms of their consequences, and they should be opposed, but doing so on the basis of a distinction between good and bad AI would not unsettle the status quo. Both serve the interests of global capitalism, and these interests are based on surveilling all areas of life, controlling labour movements and maximizing profit margins.

A similarly deficient approach would be to view the issue exclusively from the perspective of human rights, migrant rights or asylum rights and compliance with international law. While these are also important, and vested rights should be protected, on their own, they would not be sufficient to shake the status quo either. This is because surveillance capitalism, neoliberalism, postmodernism, and pro-privatization, pro-market approaches that characterize the contemporary age are not interested in complying with the norms of international law adopted in the post-war world, and have zero motivation to do so when it comes to migrants and refugees at a time when they number in the tens of millions. When we overlook this basic fact, and target people such as Trump, Johnson and Erdoğan, or the executive director of Frontex and the president of the European Commission, we lose sight of the continuity in policies, such as from Obama to Trump to Biden. This is why defending the right to asylum and the principle of non-refoulement without taking into account capitalism and the practices of surveillance capitalism would be deficient, as we would not be able to analyse the agreements made between the EU and Turkey or Libya, or those made by the US with Mexico, nor the investments of tech companies from Palantir to Microsoft, nor the weapons companies that sell the same technologies to all states for deployment on their borders. There is no contradiction between, on the one hand, the UK deciding in favour of Brexit in order to stop migrant flows and sending asylum seekers to Rwanda, and, on the other hand, the same country accepting 500,000 migrants in 2022, going beyond the pre-Brexit period. The surveillance-capitalism approach applies both to those repelled at the border and those allowed into the country.

I will not go over this again in detail, as it was discussed in Chapter 1, but the same can be applied to such questions raised in the literature as: 'Is the distinction between migrants and refugees over? Should refugees be referred to as migrants? Should the refugees stay in camps or in cities among the local population?' The observations and case studies that have given rise to these questions emphasize the contradictions and deficiencies in the implementation of human rights and asylum law. However, when we look at the issue from the perspective of the overall trajectory of capitalism, when we call attention to the practices of surveillance capitalism, neoliberal principles, pro-market biases, the rejection of public services and the leading role played by companies as stakeholders in this process, it becomes apparent that the political solutions related to these debates are still in line with the interests of local and international capital.

All of these arguments point to the need to include capitalism in the picture, which in turn requires the infrastructure to be scrutinized, that is to say, its economic foundations. The terms 'surveillance capitalism' and 'racial capitalism' are meaningful in and of themselves and emphasize the unique aspects of the issue. However, racism is not adopted for racism's sake, just as surveillance is not performed for surveillance's sake – these practices cannot be attributed to the political preferences of the capital owners. This certainly makes the issue more interesting when we focus on personalities such as Trump, Erdoğan, Peter Thiel or Elon Musk, but such an analysis will not be satisfactory. An analysis of capitalism, on the other hand, will naturally take in its infrastructure, economic foundations, production relationships, class struggles, the extraction of surplus value, profit maximization, private ownership of the means of production, monopoly power, imperialism and (neo)colonialism.

Given the focus of this book on migrants and refugees, I argue that the issue is related to global capitalism's efforts to reshape and reposition the global proletariat. For the continued extraction of surplus value, the maximization of profits and the suppression of class struggles, labour markets should be designed and shaped in line with the capitalist division of labour at an international level. One aspect of this involves the encouragement of migration. For example, there is a wealth of legal avenues available for those seeking to migrate from the Philippines to Canada to work in childcare, from Nepal to Qatar to work in construction, from Turkey to the UK to work as software developers or physicians, and from Jamaica to the UK to work as nurses. However, if you want to go to the UK from Nepal as a student, you are likely to face an arduous process, and may even face days of detention, despite the university's intervention on your behalf. If your line of work is not specified by the government, your application may be rejected despite receiving an invitation from a company, or you may be denied a tourist visa even though you are allowed to work as a physician.

Another issue relates to the control and management of the mobility of refugees, asylum seekers and people fleeing wars, violence, poverty or the climate crisis – which is not easy to predict – and dealing, in some cases, with the arrival of millions of people within a couple of weeks is no simple task. Again, in line with the international division of labour, Syrian refugees in Turkey become concentrated in the textile and other labour-intensive sectors, while Afghan refugees may be employed in agriculture. A company in Jordan can receive funding from a donor through the UN to employ Syrians and develop the gig economy, investing in fields such as catering, house cleaning and delivery, while another may use refugees as cheap labour to train algorithms in their 'artificial intelligence factory'. It is interesting here that aid organizations engage in business-oriented activities that are not directly related to the aims of the organization, such as providing mentorship and acting as accelerators with a view to creating entrepreneurs out of refugees. They also offer seed-funding support to start-ups and provide professional training and, in doing so, help to create a more skilled workforce, supporting people's participation in the labour market, mostly as workers, but sometimes as suppliers or SMEs.

Here, I underline the changes in the activities of humanitarian organizations and the relevant UN agencies (why would a humanitarian aid organization fund a start-up and provide mentorship?). In the case of irregular migration, where the legal avenues are limited, refugees and asylum seekers are expected to join the labour force as fast as possible, to become workers and start generating surplus value, and those that are able to start and manage a business are expected to join the global network of suppliers. Indeed, aid organizations work to arrange meetings between these businesses and global corporations so that they can join the supply chain.

We are talking here about millions of people on the move. It is difficult to steer legal or illegal human movements at a global level; to predict human movements; to decide which people from which nations and possessing which skills will be allowed to migrate, up to what point and where they must be stopped; and do all of this while making decisions in line with the interests of international and local/national capital. Such activities cannot be decided upon through the bilateral agreements of employment offices, ministries of labour, inspectors and experts.

It has now become possible, thanks to surveillance technologies, to overcome this difficulty to a large extent, to identify potential problems and to measure the validity and effectiveness of the measures taken. What makes it possible for these technologies to collect this data and draw conclusions is the application of the surveillance capitalism business model developed in Silicon Valley, one example of which can be seen in Turkey, where more than 4 million refugees have joined the labour force. These technologies make it difficult for refugees to go to Greece by boat, while facilitating their stay in Turkey. Some industries readily employ migrants and refugees, and are able to put hundreds of thousands of them to work without permits in the informal economy, in the textile, construction, tourism, agriculture and service sectors, along with other labour-intensive industries. The local and national capital owners in Turkey are quite happy with this arrangement, as the wages are low and the working conditions are poor (and thus, affordable), and the global capitalists based in Europe and the US who count the Turkish bourgeoisie among their suppliers, are also happy. After arriving in Turkey, a young shepherd from an Afghan village may start producing leather bags for Zara or Prada, joining the ranks of the working class. Should such migrant people ever demand their rights or engage in behaviour that would otherwise bother their bosses, they can be immediately detained and deported. This is evidence that the authorities were not unaware of the person in question when they crossed the border illegally and travelled through the country until they found a job in a factory, nor have the security forces failed to detect them. In short, their movements were practically condoned.

This raises the question of whether it would be possible or realistic to do this on a global scale. However, given the involvement of the EU and the US, along with the UK and Israel, the internal coordination among donor countries, the global footprint of the UN and its positioning at key points along migration routes together with humanitarian aid organizations, and the involvement of a small number of monopolist technology firms, it is quite possible to organize this process and sell the same products to different countries. Indeed, this is what is happening now. When we think about the border-security bureaucracies trained by the EU and UK in a large number of Asian and African countries, the border technologies sold to those states and the mechanisms established for their coordination, and consider also the similar projects, training and aid arranged or provided by the UN and relevant aid organizations in different countries, we can see that the number of actors in this field is small, and certainly fewer in number than would be expected, but very influential.

Any discussion of capitalism must first consider relations of production and class positions. Another important principle is that capitalism favours competition over markets, leading to monopolization on the one hand and regional and global wars among states on the other. Therefore, capitalist competition and market allocation cannot be considered separate from war and conflict. Throughout this book I consider the active role played by military/intelligence bureaucracies and companies in developing surveillance technologies for migration and border management, and their cooperation with tech companies. This is not, however, limited to the free movement of labour, goods and capital, and neither is the selection of migrants at the end of the story. Nor are these actors exclusively motivated by the desire to test new war equipment on migrants, given that migrants have little political power. These are all valid reasons, but when we think of the issue in connection with the phenomenon of war and the future of warfare, we must consider market competition among different powers and war becoming unavoidable.

Moving to look at the regions in need of humanitarian aid and relatedly sources of refugees and asylum seekers, we can see that they are all subject to intense competition among advanced capitalist countries. Afghanistan, Syria, Yemen, Ukraine, Somalia and many African regions are conflict zones in which different states are directly involved, providing arms to the warring parties. Competition also exists in many Latin American and Asian countries, and they also face problems such as poverty, climate crisis and mass migration due to the fierce competition over market access.

Military/intelligence companies and bureaucracies are thus in competition with one another in some countries while trying to steer people fleeing bombs or disasters and stopping them in designated places. Once these are achieved, other sectors and organizations enter the picture and turn the new arrivals into labourers, after which surveillance technologies and the accumulated knowledge and tricks of the business model of surveillance capitalism come into force for the collection of information and intelligence about people from these crisis and conflict zones and manipulate and attract societies using different methods.

Our focus in the field of migration is people who cross or attempt to cross borders, although I would like to emphasize that the smart border technologies tested on migrants also serve to militarize the regions in question for actual or potential wars, potentially putting peace at risk and spreading conflict to new regions. This, therefore, is not just about testing new ideas or products, as these products are actively used in the struggle for hegemony and market dominance in actual wars or in conflicts that have not escalated to physical violence. For example, Turkey and Greece deploy the latest technologies on their borders using their own resources and EU funds. While the stated goal is to stop migration flows, the underlying issue is clearly the continuation, using non-violent means, of the regional rivalry and the fight for hegemony between Turkey and Greece. As these technologies are also, or mainly, intended for intelligence gathering from the other country, the result is the further militarization of the Aegean Sea. We are thus confronted with the question of whether the tragedy faced by migrant people in the Aegean is a result of anti-migrant attitudes on the part of the two countries or an outcome of the regional tension and rivalry over access to markets and international waters. Focusing exclusively on migrants would fail to provide a satisfactory answer to this question.

On the other hand, projects to issue ID cards to the more than 1 billion people with no official IDs, open bank accounts for the 'unbanked', and provide SIM cards and smartphones to everyone mean that the large corporations that invest in these fields gain access to new markets, find new customers and collect data from millions of people who were previously out of reach. These projects, as well as efforts to establish databases for the storage of the biometric data of the hundreds of millions of people and the practice of collecting data on all aspects of people's lives when they apply for visas and permits, are motivated by the huge appetite for data of surveillance capitalism and intelligence agencies.

Therefore, when evaluating the policies and tech products related to migrants and asylum seekers, including products considered to be pro-migrant, we have no choice but to consider the overall trajectory of capitalism, market allocation, the positioning of people as labourers and consumers, the process of monopolization and the violent aspects of this process.

Post-consent era?

I have discussed the issues of privacy and consent in each chapter, dealing with different technologies and, in addition to underlining their importance, I have discussed the significance of the erosion of these principles through surveillance capitalism. It is clear that finding effective solutions to these issues is no easy task. Once it is decided, for example, that cash transfers to a community in Kenya will be carried out using blockchain, or that a large database will be provided by a company like Palantir, or that AI algorithms will be created for another project, the preparation required and the costs involved are overwhelming, and it is very difficult for all those who take part in these projects to understand these technologies. It may also be difficult for the beneficiaries to participate in the decision-making process or to reject the aid and take care of their own problems if they have some concerns.

This gives rise to such questions as, 'Are issues of privacy and consent even relevant anymore?' Given that these concerns also exist in places with high privacy standards and strong legal enforcement, such as the EU and US, it is impossible to be reassured by the legal protection provided in developing countries. Should we then give up on these principles?

Are we living in a post-consent society? Are we not really affected, on a personal level, by the huge amounts of data that is collected and analysed? For example, is it a meaningful act to accept cookies when we visit websites? Does anyone read the conditions? Is it really such a big deal to give up some privacy in return for the services received? Have the related issues of consent and privacy been turned into problems by lawmakers who fail to grasp the reality, dynamism and speed of the digital world? Are we talking about laws that were rendered irrelevant because they preceded the recent advances in technology? Wouldn't we be better off if we were more realistic and practically minded and stopped insisting on consent?

In response to these questions, one concrete proposal made by people who have adopted such an approach is that the issue of privacy should be taken into account by companies at the time of product development, beginning in the design stage. Being part of the process from design onwards is one of the main proposals of those who adopt a reformist approach to technological progress. According to this approach, if the data sets used to train the algorithm do a poor job of representing society, if they are marred with inequalities and discrimination, then area experts and relevant NGOs should take part and even supervise the entire process, from the design stage to the training of the algorithms and developing solutions to any problems that may arise. In another example, if experts from a labour union take part in and supervise the process of designing and coding the human resources algorithm to be used in a workplace, they may be able to prevent the algorithm from being unfair to the workers in the future. As a 'practical and realist' solution, when the issue involves vulnerable communities such as refugees, for example, the companies that develop the products can work together with UN agencies and aid organizations to implement the project, from product design to coding, and solve issues of privacy and consent by taking the unique situation of refugees into account. In this case, they would no longer be required to inform individual refugees or migrants, or obtain their consent.

This approach merits discussion, and it would certainly be useful to research various practical solutions. Based on the issues discussed in this book, however, I believe the concept of a post-consent society to be unacceptable. When we think of the issue in terms of surveillance capitalism and the related concepts of deregulation, neoliberalism and market fundamentalism, it amounts to a call for advocates of migrant and refugee rights to give up an important principle in the name of being practically minded and realistic. It reflects a desire for less regulation, fewer rules, more dominance by private companies, less control, less accountability, more data collection and more speculative projects being implemented. Some things may be *difficult* to achieve in real life, but that does not make them impossible or unattainable. Giving up the fight for principles, accepting the status quo and leaving one's concerns aside may be acceptable for organizations, but for people on the ground – the 'beneficiaries' – it amounts to yet another usurpation of their rights and represents the reproduction of a colonialist, top-down approach that denies people's agency.

For people who are concerned about and would like to limit the dominance of surveillance capitalism in the fields of humanitarian aid, migration, asylum-seeking and borders, consent and privacy should be indispensable principles. At the very least, asking people whether they give consent and making public declarations of respect for privacy would be a good start. This would make it possible to ask whether promises are being kept, contributing to accountability and opening the way for further demands. Of course, basing our opposition on privacy and consent alone would not suffice, and so, looking at the issue from the perspective of surveillance capitalism and trying to understand its role in the overall trajectory of capitalism would make these debates more meaningful.

In short, it is important to expose the inadequacies of the status quo, rather than protecting it or compromising with it, to show that the issue is not just about migrants' rights, and that migrants are not accorded even the slightest respect. On the other hand, saying, 'Let's not bother with obtaining the consent of millions of people. Those times are over; the solution is to have products designed that respect the principles' is an approach that expects the solution from the creators of the problem, leaving the initiative to companies and reinforcing the hegemony of surveillance capitalism. UN agencies and humanitarian bodies are already forced to work with companies as they lack the required technical teams. If, on top of that, they were to transfer their privacy and consent responsibilities to companies that they do not control, the deregulation and erosion of rules would be reinforced.

This reflects the point made in Chapter 4, suggesting that the solutions to privacy and cryptography could be offered by companies that adopt the surveillance capitalism business model, such as Microsoft. Arguing that they are in the best position to resolve the problems that they have created, and proposing to resolve political problems with better technological products, they treat the issues as an opportunity to test yet more technological products on migrants and refugees. By using migrant people as test subjects, they see whether the cryptography actually works, and whether there are any system errors or malfunctions, while the risk of migrant people being hurt in the process is not even considered. When we leave the initiative to private companies and the market, issues such as engaging in a lively political debate, coming up with creative solutions and ensuring the participation of migrant people in the process are disregarded.

What if the system malfunctions?

As the dominance of corporations in the fields of humanitarian aid and migration management becomes consolidated and the public sector takes a back seat, issues such as consent and privacy can be expected to be ignored in favour of large-scale, speculative projects, while the potential negative impacts of these projects and the technological solutions offered, even when successful, are not discussed. This is partly because a significant number of academics who conduct research in these fields and take a critical position are unfamiliar with technological advances and their technical aspects. As mentioned in Chapter 3, however, when virtual walls are erected on borders or humanitarian aid is provided, and migration management is carried out based on analyses of data collected using the most advanced technologies, it is quite possible for these systems to malfunction in everyday life.

When that happens, rejecting an immigrant's or refugee's application due to a system error can potentially have devastating consequences, and we know that there is almost no remedy for such situations. For example, the Eurodac

2020 Annual Report refers to errors in the analysis of asylum applications as 'wrong hits' and 'missed hits', stating: 'Those delays were responsible for producing 1,300 wrong hits, most of these (74% of the wrong hits) were due to data registered by Hungary. ... Those delays were responsible for producing 67 missed hits. Half of them – 35 missed hits – related to data that were submitted late by Spain' (Eurodac, 2021).[1]

These errors and delays can have critical, life-or-death consequences for those involved. In the absence of accountability, the report nonchalantly states that there is a problem with the applications of about 1,400 people, as if it were an ordinary technical issue. It is difficult to even imagine the psychological and social pressure the applicants must have endured as a result. The report fails to detail what happened to the approximately 1,400 people mentioned, and there is no way of knowing what other problems went undetected.

On the one hand, there may be system errors related to the applicants, as well as issues related directly to border security. By deploying technological products at their borders, states attempt to create virtual walls and make their borders more secure (Vukov and Sheller, 2013). As mentioned in Chapter 3, border security would be at serious risk if these systems were hacked by hostile countries or malfunctioned. This is another example of the need to question the role played by corporations in these areas, based on their lobbying power, technical superiority and financial means.

Alternative tech?

A significant proportion of the work on digital technologies in migration studies is devoted to how migrants and refugees use these products. These studies, most of which are ethnographic in nature, document the importance of smartphones, WhatsApp and Facebook groups, Telegram and TikTok posts, and map and translation apps for people who use digital technologies to plan their migration, to gain an understanding of the process of collecting documents and submitting an application if legal routes are to be used, to contact smugglers if irregular routes are to be used, to learn about people and organizations that provide help, and to learn about the legal, social and economic issues they are likely to face during their journey and after their arrival. This is why being able to charge phones and access the internet at borders, in airports and in refugee camps is vital (Gillespie et al, 2016).

There are also studies in the literature detailing the social and psychological problems experienced by 'Skype mothers', women who come to Europe from the Philippines for domestic work, and who meet regularly with their children and families via Skype. Moreover, there are studies on engagement with the diaspora examining how digital technologies bring diasporas together and how they are used to organize political and financial campaigns in the country of origin. These campaigns may be political in nature, offering support against authoritarian governments back home or calling for the recognition of the right of nations for self-determination, or may focus on humanitarian activities and economic support after natural disasters. Digital technologies make it quite possible for campaigns to be organized and carried out quickly, reliably and with mass participation.

Such technologies can also be used by pro-migrant and humanitarian aid bodies in ways that run counter to the examples mentioned in previous chapters. Alternative mobile apps are developed to allow people on the move to avoid Big Tech companies, circumvent surveillance or migrate without the need for smugglers. Similarly, NGOs such as Sea-Watch utilize open-access data sets and commercial data from satellites to identify migrants attempting boat crossings, and endeavour to arrive before the coastguard does, supporting their safe landing and asylum applications (Interaction, 2003; Hernandez and Roberts, 2018; Kahng et al, 2018; Kanter and Fine, 2018; Kaurin, 2019).

It is also known that migrants and refugees have created groups on social media platforms in which they rate smugglers, share their experiences and make recommendations in the style of, say, TripAdvisor. As these sites are known to be monitored by security and intelligence agencies, migrant people take certain precautions when communicating with one another and negotiating with smugglers. Previous studies have reported that, although such digital communications are common, online meetings and plans are made only with real-life acquaintances or relatives, and the groups change their names frequently.

As discussed elsewhere in the book, there are significant digital inequalities among migrant people, based on age, gender, class and education, and those unable to access information through digital means are obviously more disadvantaged. For example, people may learn that they do not need the assistance of smugglers after crossing into Turkey from Iran, as they can head directly to Ankara and apply to the UNHCR. Such information can be garnered from the internet and the support gained through apps, so the difficulties they face are qualitatively different from those faced by people who do not have access to such information and support.

Technological products are commodities that have use- and exchange-value. As such, they can be used for the benefit of migrants and humanity in general, or as tools of oppression and exploitation. The important thing is to distinguish between those who create value and those who make a profit, and the fact that we now know a lot about the qualitative and quantitative impact of the business model of surveillance capitalism on monopolization and the overall trajectory of capitalism is an important advantage in this regard (Zuboff, 2019).

We should, however, avoid romanticizing the use of digital technologies by migrants. The system, equipped under the business model of surveillance capitalism, plays a dominant role in steering migration movements through surveillance and data analysis. It is not an easy task for migrant people to take all these risks and to distinguish between right and wrong, friend and foe, or disinformation and accurate information, which, ultimately, is of secondary importance anyway.

Obviously, for someone who has a residence permit or a talent visa and who works as an academic in a university or as a financial specialist in a bank, the risks are lower and are not life-threatening. However, in a country like the UK, for example, where there is a significant wage gap between people of different classes and ethnicities, even between those doing the same job, and where online misinformation and manipulation campaigns influence election outcomes, as was arguably the case during the Brexit referendum, migrants receive migrant-specific economic and political advertising messages (for remittances, citizenship and so on), and are more likely to be negatively affected by surveillance than their colleagues who are citizens. Add to this the requirement to provide biometric data and information about all aspects of one's life when applying for a residence permit, which must be updated during renewal applications, and it becomes clear that migrant people would be in a more precarious position if they ever got into trouble.

Moving forward

I am aware of the difficulty in making any concrete recommendations given the scope of this book; however, I would like to end with a few thoughts on further research and discussions.

First of all, it is very important to ensure that political problems related to the fields of migration and humanitarian aid are addressed through lively political debates with the participation of all relevant parties, as should be the case in all political issues. The technological products in question allow processes of deliberation and negotiation of this sort, and the participation of migrants in this process is crucial. Solving problems with more advanced technological products leads to greater dependence on monopolist tech companies and intelligence/security bureaucracies. As I have argued earlier, when there are privacy concerns related to a project, the solution is not necessarily to develop a product with stronger cryptographic features.

It is, of course, impossible to reject the use of digital technologies in these fields, although we can reject surveillance capitalism and its inherent business model. Applying the principle of 'no harm' can exclude companies that adopt and profit from the surveillance capitalism business model and that train and perfect their algorithms through data-analysis approaches. Similarly, the principle of 'neutrality' can be invoked to ban tech companies from working with arms manufacturers and military/security agencies when investing in the fields of humanitarian aid and migration (ICRC and Privacy International, 2018).

If an organization working in this field undertakes activities that benefit hundreds of thousands or millions of people in multiple countries and is expected to do so for many years, and when the use of digital solutions is a must, they should be required to have a sufficiently qualified technical team and to develop technological products in house and in line with the requirements of their specific field. This would thus make it possible to choose products that are easier to understand for beneficiaries and fieldworkers, to come up with new approaches to privacy and consent issues, and to develop alternative solutions for when problems arise and for cases where consent is not given, based perhaps on feedback from the beneficiaries and fieldworkers. This is not unprecedented, as some NGOs have been using their own software for years in the service of millions of migrant people, and their experiences may be constructive in this regard.

Most importantly, however, this is an issue that concerns not only migrants, refugees, organizations working in this field and academics working on migration. The effects of surveillance technologies developed and tested in this field will likely be felt by all of humanity in the near future. Investments and pilot projects in these fields also reinforce militarism and warmongering. In this respect, researchers who study surveillance capitalism and monopolist tech companies and their relationships with other sectors, and the future of warfare and the future of work would also benefit from a study of the technological projects being implemented in the fields of migration and border management.

There is an obvious reason for this: it is impossible to mount an effective opposition without making capitalism part of the discussion and without recognizing the dystopian future that contemporary capitalism envisions for humanity through its various practices. The fields of migration and border management and humanitarian aid are among the lowest hanging fruit for these corporations and have the potential to affect everyone. This book aims to have contributed to the ongoing debates in this regard, and to invite further discussion on the future envisioned for humanity by capitalism.

Notes

Introduction
1. Exchange-value is the quantitative aspect of value, as opposed to 'use-value', which is the qualitative aspect of value and constitutes the substratum of the price of a commodity (marxists.org/glossary/terms/e/x.htm).

Chapter 1
1. The Ukraine case is certainly interesting because perhaps *more than any other group of refugees*, Ukrainian refugees received specialized psychological support and also free movement to return to Ukraine (and then return to protection-granting countries). But this does not change the 'normalcy' approach to send them to workplaces.

    The more devastating issue in the Ukrainian case is that third-country nationals were barred from even leaving Ukraine, completely forgotten in terms of support, and then immediately shipped back to their country of origin without regard for their specific situations, residencies or home contexts.
2. Here, I do not elaborate on rentier capitalism in relation to debates on digital feudalism. I do not deny that tech corporations create digital feudal relations, but they are similar to other monopolistic corporations, benefitting from monopoly rent and providing platforms (Morozov, 2022; Lenin, 2008 [1917]).

Conclusion
1. A footnote on page 19 of the report defines these terms, but in a tiny type size:

    An example of a so-called 'wrong hit': a third-country national lodges an international protection application in Member State A, whose authorities take the person's fingerprints. While those fingerprints are awaiting transmission to the Eurodac (Category 1 transaction), the same person could go to Member State B and lodge another application. If Member State B sends the fingerprint data before Member State A, the fingerprint data sent by Member State A would be registered in the Eurodac later than the fingerprint data sent by Member State B. This would result in a hit from the data sent by Member State B against the data sent by Member State A (a wrong hit). Member State B would therefore be deemed responsible instead of Member State A, where the application was first lodged.

    An example of a so-called 'missed hit': a third-country national or stateless person is apprehended in connection with an irregular border crossing and the person's fingerprints are taken by the authorities of Member State A. While those fingerprints are awaiting transmission to the Eurodac (Category 2 transaction), the same person could go to Member State B and lodge an application for international protection. At that time, fingerprints are taken by the authorities of Member State B. If Member State B sends the fingerprint data (Category 1 transaction) before Member State A, the Eurodac would register this as a Category 1 transaction, and Member State B would have to handle the application instead of Member State A. When the Category 2 transaction arrives later, a hit will be missed, because Category 2 data are not searchable.

References

Achiume, T. (2020) 'Racial discrimination and emerging digital technologies: report of the Special Rapporteur on Contemporary Forms of Racism, Racial Discrimination, Xenophobia and Related Intolerance', 18 June, Geneva: United Nations, digitallibrary.un.org/record/3879751

Akhmetova, R. and Harris, E. (2021) 'Politics of technology: the use of artificial intelligence by US and Canadian immigration agencies and their impacts on human rights', in E. E. Korkmaz (ed) *Digital Identity, Virtual Borders and Social Media*, Cheltenham: Edward Elgar Publishing.

Akkerman, M. (2016) 'Border wars: the arms dealers profiting from Europe's refugee tragedy', Transnational Institute and Stop Wapenhandel.

Akkerman, M. (2021) *Financing Border War: The border industry, its financiers and human rights*, Transnational Institute and Stop Wapenhandel.

Albahari, M. (2015) *Crimes of Peace: Mediterranean migrations at the world's deadliest border*, Philadelphia: University of Pennsylvania Press.

Amin, S. (1974) 'Accumulation and development: a theoretical model', *Review of African Political Economy*, 1(1): 9–26.

Anderson, B. (2000) *Doing the Dirty Work: The global politics of domestic labour*, London: Zed Books.

Andersson, R. (2022) 'The bioeconomy and the birth of a "new anthropology"', *Cultural Anthropology*, 37 (1): 37–44.

Angwin, J., Larson, J., Mattu, S. and Kirchner, L. (2016) 'Machine bias', *ProPublica*, propublica.org/article/machine-bias-risk-assessments-in-criminal-sentencing

Arango, J. (2000) 'Explaining migration: a critical view', *International Social Science Journal*, 52 (165): 283–296.

Au, Y. (2021) 'Surveillance as a service: the European AI-assisted mass surveillance marketplace', *Oxford Commission on AI and Good Governance*, Oxford Internet Institute, November.

Aus, J. P. (2006) 'Eurodac: a solution looking for a problem?' *(Working Paper No. 9)*, Oslo: Centre for European Studies, University of Oslo.

Azieki, M., Boyce, G. and Todd, M. (2021) *Smart Borders or a Humane World?* The Immigrant Defense Project's Surveillance, Tech & Immigration Policing Project, and the Transnational Institute.

Bankston, J. (2021) 'Migration and smuggling across virtual borders: a European Union case study of internet governance and immigration politics', in E. E. Korkmaz (ed) *Digital Identity, Virtual Borders and Social Media*, Cheltenham: Edward Elgar Publishing.

Bansak, K., Ferwerda, J., Hainmueller, J., Dillon, A., Hangartner, D., Lawrence, D. et al (2018) 'Improving refugee integration through data-driven algorithmic assignment', *Science*, 359 (6373): 325–329.

Barbrook, R. and Cameron, A. (1996) 'The Californian ideology', *Science as Culture*, 6 (1): 44–72.

Bauman, Z. (1998) *Postmodernity and its Discontents*, Cambridge: Polity Press/Blackwell.

Beduschi, A. (2017) 'The big data of international migration: opportunities and challenges for states under international human rights law', *Georgetown Journal of International Law*, 49 (3): 981–1017.

Beniger, J. (2009) *The Control Revolution: Technological and economic origins of the information society*, Cambridge, MA: Harvard University Press.

Benjamin, R. (2016) 'Catching our breath: Critical Race STS and the carceral imagination', *Engaging Science, Technology, and Society*, 2: 145–156.

Benner, T. (2020) 'Britain knows it's selling out its national security to Huawei', *Foreign Policy*, 31 January.

Benson, T. (2015) 'Ways we must regulate drones at the us border', *Wired*, wired.com/2015/05/drones-at-the-border

Bigo, D. (2002) 'Security and immigration: toward a critique of the governmentality of unease', *Alternatives*, 27 (1): 63–92.

Bigo, D. (2007) 'Detention of foreigners, states of exception, and the social practices of control of the Banopticon', in P. Rajaram and C. Grundy-Warr (eds) *Borderscapes: Hidden geographies and politics at territory's edge*, Minneapolis: University of Minnesota Press, pp 3–33.

Bigo, D. and Guild, E. (eds) (2005) *Controlling Frontiers: Free movement into and within Europe*, Aldershot: Ashgate.

Bircan, T. (2022) 'Remote sensing data for migration research', in A. A. Salah, E. E. Korkmaz and T. Bircan (eds) *Data Science for Migration and Mobility*, Oxford: Oxford University Press.

Bircan, T. and Korkmaz, E. E. (2021) 'Big data for whose sake? Governing migration through artificial intelligence', *Nature Humanities & Social Sciences Communication*, October.

Bird, G. (2022) 'Smart borders: EU border agency Frontex accused of covering up human rights violations in Greece – the allegations explained', *Conversation UK*, 17 October.

Bozçağa, T. and Cansuna, A. (2022) 'Combining mobile call data and satellite imaging for human mobility', in A. A. Salah, E. E. Korkmaz and T. Bircan (eds) *Data Science for Migration and Mobility*, Oxford: Oxford University Press.

Brinham, N. (2019) 'When identity documents and registration produce exclusion: lessons from Rohingya experiences in Myanmar', Middle East Centre blog, LSE, blogs.lse.ac.uk/mec/2019/05/10/when-identity-documents-and-registration-produce-exclusion-lessons-from-rohingya-experiences-in-myanmar

Broeders, D. (2007) 'The new digital borders of Europe: EU databases and the surveillance of irregular migrants', *International Sociology*, 22 (1): 71–92.

Browne, S. (2015) *Dark Matters: On the surveillance of blackness*, Durham, NC: Duke University Press.

Burns, M. (2015) 'Leaked Palantir doc reveals uses, specific functions and key clients', *Techcrunch.Com*, techcrunch.com/2015/01/11/leaked-palantir-doc-reveals-uses-specific-functions-and-key-clients

Burns, R. (2015) 'Rethinking big data in digital humanitarianism: practices, epistemologies, and social relations', *GeoJournal*, 80 (4): 477–490.

Cabanes, J. V. A. and Acedera, K. A. F. (2012) 'Of mobile phones and mother-fathers: calls, text messages, and conjugal power relations in mother-away Filipino families', *New Media and Society*, 14 (6): 916–930.

Calderón, C. A., Amores, J. J. and Stnek, M. (2022) 'Machine learning and synthetic populations to predict social acceptance of asylum seekers in European regions', in A. A. Salah, E. E. Korkmaz and T. Bircan (eds) *Data Science for Migration and Mobility*, Oxford: Oxford University Press.

Castles, S. (2010) 'Understanding global migration: a social transformation perspective', *Journal of Ethnic and Migration Studies*, 36 (10): 1565–1586.

Castles, S. and Wise, R. D. (eds) (2007) *Migration and Development: Perspectives from the South*, Geneva: IOM.

Cheesman, M. and Slavin, A. (2021) 'Self-sovereign identity and forced migration: slippery terms and the refugee data apparatus', in E. E. Korkmaz (ed) *Digital Identity, Virtual Borders and Social Media*, Cheltenham: Edward Elgar Publishing.

Cheney-Lippold, J. (2011) 'A new algorithmic identity: soft biopolitics and the modulation of control', *Theory, Culture & Society*, 28 (6): 164–181.

Christophers, B. (2022) *Rentier Capitalism: Who Owns the Economy, and Who Pays for It?*, New York: Verso.

Cohen, N. (2017) 'The libertarian logic of Peter Thiel', Wired, wired.com/story/the-libertarian-logic-of-peter-thiel

Constantin, L. (2016) 'Armies of hacked IoT devices launch unprecedented DDoS attacks', CSO, csoonline.com/article/3124344/armies-of-hacked-iot-devices-launch-unprecedented-ddos-attacks.html

Coppi, G. and Fast, L. (2019) 'Blockchain and distributed ledger technologies in the humanitarian sector', commissioned report, Humanitarian Policy Group at the Overseas Development Institute, pp 1–37.

Coppi, G., Jiminez, R. M. and Kyriazi, S. (2021) 'Explicability of humanitarian AI: a matter of principles', *Journal of International Humanitarian Action*, 6 (19), https://doi.org/10.1186/s41018-021-00096-6

Cottray, O. and Larrauri, H. P. (2017) 'Technology at the service of peace', SIPRI blog, 26 April, sipri.org/commentary/blog/2017/technology-service-peace

Cresswell, T. (2006) *On the Move: Mobility in the Modern Western World*, London: Routledge.

Crisp, J. (2018) 'Beware the notion that better data lead to better outcomes for refugees and migrants', Chatham House, 9 March.

Csernatori, R. (2018) 'Constructing the EU's high-tech borders: Frontex and dual-use drones for border management', *European Security*, 27 (2): 175–200.

Cummings, M. L., Roff, H. M., Cukier, K., Parakilas, J. and Bryce, H. (2018) 'Artificial intelligence and international affairs: disruption anticipated', Royal Institute of International Affairs, London.

Daniels, E. (2018) 'Lie-detecting computer kiosks equipped with artificial intelligence look like the future of border security', CNBC, 15 May, https://www.cnbc.com/2018/05/15/lie-detectors-with-artificial-intelligence-are-future-of-border-security.html

d'Appollonia, A. C. (2012) *Frontiers of Fear: Immigration and Insecurity in the United States and Europe*, Ithaca: Cornell University Press.

Dave, A. (2017) 'Review of *Digital Humanitarians* by Patrick Meier', *Journal of Bioethical Inquiry*, 14 (4): 567–569.

Davies, R. (2021) '"Conditioning an entire society": the rise of biometric data technology', *The Guardian*, 26 October, https://www.theguardian.com/technology/2021/oct/26/conditioning-an-entire-society-the-rise-of-biometric-data-technology

De Genova, N. (2013) 'Spectacles of migrant "illegality": the scene of exclusion, the obscene of inclusion', *Ethnic and Racial Studies*, 36 (7): 1180–1198.

Delcan, P. (2019) 'Meet America's newest military giant: Amazon', *MIT Technology Review*, November.

De Leon, J. (2015) *The Land of Open Graves: Living and dying on the migrant trail*, Berkeley: University of California Press.

Deridder, M., Pelckmans, L. and Ward, E. (2020) 'Reversing the gaze: West Africa performing the EU migration-development-security nexus', *Anthropologie & Développement*, 51: 9–32.

Duffield, M. (2012) 'Challenging environments: danger, resilience and the aid industry', *Security Dialogue*, 43 (5): 475–492.

Dumbrava, C. (2021) 'Artificial intelligence at EU borders: overview of applications and key issues', European Parliamentary Research Service.

Escobar, A. (1995) *Encountering Development: The making and unmaking of the Third World*, Princeton: Princeton University Press.

REFERENCES

Eurodac (2021) *Eurodac* 2020 Annual Report, https://www.eulisa.europa.eu/Publications/Reports/Eurodac%20AR%202020.pdf

FATF/OECD (2020) *Guidance on Digital Identity: Executive summary*, pp 5–12, fatf-gafi.org/media/fatf/documents/recommendations/pdfs/Guidance-on-Digital-Identity-Executive-Summary.pdf

Feldstein, S. (2019) *The Global Expansion of AI Surveillance*, Carnegie Endowment for International Peace.

Feuer, W. (2020) 'Palantir CEO Alex Karp defends his company's relationship with government agencies', CNBC, 23 January, cnbc.com/2020/01/23/palantir-ceo-alex-karp-defends-his-companys-work-for-the-government.html

Floridi, L. (2007) 'A look into the future impact of ICT on our lives', *Information Society*, 23 (1): 59–64.

Francisco, V. (2018) *The Labor of Care: Filipina migrants and transnational families in the digital age*, Champaign: University of Illinois Press.

Franco, M. (2019) 'Democrats want a "smart wall". That's trump's wall by another name', *The Guardian*, theguardian.com/commentisfree/2019/feb/14/democrats-wall-border-trump-security

Fuchs, C. (2019) *Rereading Marx in the Age of Digital Capitalism*, London: Pluto Press.

Fussell, S. (2019) 'The increase in drones used for border surveillance', *The Atlantic,* theatlantic.com/technology/archive/2019/10/increase-drones-used-border-surveillance/599077

Gardner, A. (2011) *City of Strangers: Gulf migration and the Indian community in Bahrain*, Ithaca: Cornell University Press.

Gazzotti, L. (2021) '(Un)making illegality: border control, racialized bodies and differential regimes of illegality in Morocco', *The Sociological Review*, 69 (2): 277–295.

Ghaffary, S. (2019) 'The "smarter" wall: how drones, sensors, and AI are patrolling the border', Vox, vox.com/recode/2019/5/16/18511583/smart-border-walldrones-sensors-ai

Gillespie, M., Ampofo, L., Cheesman, M. and Fath, B. (2016) *Mapping Refugee Media Journeys: Smartphones and social media networks*, technical report, The Open University and France Medias Monde.

Gilroy, P. (1993) *The Black Atlantic: Modernity and Double-consciousness,* Cambridge, MA: Harvard University Press.

Greenwood, F. (2019) 'Why humanitarians are worried about Palantir's partnership with the U.N.', Slate, slate.com/technology/2019/02/palantir-un-world-food-programme-data-humanitarians.html

GSMA (2018) 'Using mobile technology to provide functional identities', GSMA blog, 22 January, gsma.com/mobilefordevelopment/blog-2/using-mobile-technology-provide-functional-identities

Gürkan, M., Bozkaya, B. and Balcısoy, S. (2022) 'Financial datasets: leveraging transactional big data in mobility and migration studies', in A. A. Salah, E. E. Korkmaz and T. Bircan (eds) *Data Science for Migration and Mobility*, Oxford: Oxford University Press.

Hall, A. E. (2017) 'Decisions at the data border: discretion, discernment and security', *Security Dialogue*, pp 488–504, https://doi.org/10.1177/09670 10617733668

Hampshire, J. and Broeders, D. (2010) *The Digitalization of European Borders and Migration Controls: Migration to Europe in the digital age*, MEDiA report on Work Package 2, MEDiA meeting at Koc University, Istanbul, 9–10 April.

Hatmaker, T. (2019) 'Palantir wins $800 million contract to build the US Army's next battlefield software system', TechCrunch, techcrunch.com/2019/03/27/palantir-army-contract-dcgs-a

Hernandez, K. and Roberts, T. (2018) 'Leaving no one behind in a digital world', technical report, *The K4D Emerging Issues Report Series*, Institute of Development Studies.

HIP (2020) 'Layering digital id on top of traditional data management', hiplatform.org/blog/2020/5/20/layering-digital-id-on-top-of-traditional-data-management

Holloway, K., Al Masri, R. and Abu Yahia, A. (2021) 'Digital identity, biometrics and inclusion in humanitarian responses to refugee crises', ODI working paper, 6 October, https://odi.org/en/publications/digital-identity-biometrics-and-inclusion-in-humanitarian-responses-to-refugee-crises/

Hosein, G. and Nyst, C. (2013) *Exploration of How Development and Humanitarian Aid Initiatives Are Enabling Surveillance in Developing Countries*, Privacy International, https://ssrn.com/abstract=2326229

ICRC and Privacy International (2018) *The Humanitarian Metadata Problem: 'Doing no harm' in the digital era*, technical report, privacyinternational.org/report/2509/humanitarian-metadata- problem-doing-no-harm-digital-era

Interaction (2003) 'Diagnostic tool and guidance on the interaction between field protection clusters and UN missions', technical report, Global Protection Cluster, globalprotectioncluster.org/assets/files/toolsandguidance/InterActionGuideIncorporatingProtection2003EN.pdf

Jacobsen, K. L. (2015) *The Politics of Humanitarian Technology: Good intentions, unintended consequences and insecurity*, Abingdon: Routledge.

Joque, J. (2022) *Revolutionary Mathematics: Artificial intelligence, statistics and the logic of capitalism*, New York: Verso.

Juskalian, R. (2018) 'Inside the Jordan refugee camp that runs on blockchain', *MIT Technology Review*, 12 April, technologyreview.com/s/610806/inside-the-jordan-refugee-camp-that-runs-on-blockchain

Kahng, A., Mackenzie, S. and Procaccia, A. D. (2018) *Liquid Democracy: An algorithmic perspective*, in 32nd AAAI Conference on Artificial Intelligence.

Kanter, B. and Fine, A. (2018) 'Civil society in the age of automation: understanding the benefits and risks of artificial intelligence, machine learning, and bots', *Alliance for Peacebuilding and Toda Peace Institute Policy Brief*, 26, November, https://static1.squarespace.com/static/5db70e83fc0a966cf4cc42ea/t/5ea9d760fa1d4b1c80c3e4e0/1588189024552/t-pb-26_kanter-and-fine.civil-society-in-the-age-of-automation.pdf

Kaurin, D. (2019) 'Data protection and digital agency for refugees', technical report 12, World Refugee Council Research Paper.

Kello, L. (2019) *The Virtual Weapon and International Order*, New Haven and London: Yale University Press.

Kelly, P. (2022) *Artificial Intelligence: Report of the Standing Committee on Access to Information, Privacy and Ethics*, October, House of Commons Canada.

Keung, N. (2017) 'Canadian immigration applications could soon be assessed by computers', *Toronto Star*, thestar.com/news/immigration/2017/01/05/immigration-applications-could-soon-be-assessed-by-computers.html

Korkmaz, E. E. (2017) 'How do Syrian refugee workers challenge supply chain management in the Turkish garment industry?', *International Migration Institute Working Paper Series*, paper 133.

Korkmaz, E. E. (2018) 'How do Turkey-origin immigrant workers in Germany represent themselves through trade unions and works councils?', *Economic and Industrial Democracy*, 42 (3): 716–736.

Kolkman, D. (2020) '"F**k the algorithm"? What the world can learn from the UK's A-level grading fiasco', LSE Blog, https://blogs.lse.ac.uk/impactofsocialsciences/2020/08/26/fk-the-algorithm-what-the-world-can-learn-from-the-uks-a-level-grading-fiasco/

Kuster, B. and Tsianos, V. (2016) 'How to liquefy a body on the move: Eurodac and the making of the European digital border', in R. Bossong and H. Carrapico (eds) *EU Borders and Shifting Internal Security*, Cham: Springer.

Larsson, S. (2020) 'The civil paradox: Swedish arms production and export and the role of emerging security technologies', *International Journal of Migration and Border Studies*, 6(1/2).

Latonero, M. (2019) 'Stop surveillance humanitarianism', *The New York Times*, 11 July.

Latonero, M., Poole, D. and Berens, J. (2018) 'Refugee connectivity: a survey of mobile phones, mental health, and privacy at a Syrian refugee camp in Greece', *Nature*, nature.com/articles/d41586-019-01679-5

Lenin, V. I. (2008 [1917]) *Imperialism: The highest stage of capitalism*, Broadway, NSW: Resistance Books.

Letouzé, E., Meier, P. and Vinck, P. (2013) 'Big data for conflict prevention: new oil and old fires', in F. Mancini (ed) *New Technology and the Prevention of Violence and Conflict*, New York: International Peace Institute.

Lighthouse Reports (2022) Europe's Black Sites, lighthousereports.nl/investigation/europes-black-sites

Lyon, D. (2009) *Identifying Citizens: ID cards as surveillance*, Cambridge: Polity Press.

Macias, L. (2019) 'Entre contrôle et protection: ce que les technologies de l'information et de la communication font au camp de réfugiés', *Communications*, 104: 107–117.

Madianou, M. (2019) 'Technocolonialism: digital innovation and data practices in the humanitarian response to refugee crises', *Social Media + Society*, 5 (3): 1–13.

Massey, D. S. (2015) 'A missing element in migration theories', *Migration Letters*, 12 (3): 279–299.

Mastercard (2019) *Digital Identity: Restoring trust in a digital world*, mastercard, us/content/dam/mccom/en-us/issuers/digital-identity/digital-identity-restoring-trust-in-a-digital-world-final-share-corrected.pdf

Maxmen, A. (2019) 'Can tracking people through phone-call data improve lives?', *Nature* (online), 29 May, nature.com/articles/d41586-019-01679-5

Mayer-Schönberger, V. and Ramme, T. (2018) *Reinventing Capitalism in the Age of Big Data*, London: John Murray Press.

Mbembe, A. (2003) 'Necropolitics', *Public Culture*, 15 (1): 11–40.

Meier, P. (2011) 'Do "Liberation Technologies" change the balance of power between repressive states and civil society?' unpublished PhD thesis, Fletcher School of Law and Diplomacy.

Meier, P. (2015) *Digital Humanitarians: How big data is changing the face of humanitarian response*, Abingdon: Routledge.

Mercy Corps (2016) 'How technology is affecting the refugee crisis: Afghanistan, Greece, Iraq, Jordan, Syria', 9 June, mercycorps.org/blog/technology-refugee-crisis

Mijente (2019) *The War Against Immigrants: Trump's tech tools powered by Palantir*, mijente.net/wp-content/uploads/2019/08/Mijente-The-War-Against-Immigrants_-Trumps-Tech-Tools-Powered-by-Palantir_.pdf

Mijente (2022) *HART Attack: How DHS's massive biometrics database will supercharge surveillance and threaten rights immigration defense project*, with Just Futures Law and Immigrant Defense Project, immigrantdefenseproject.org/wp-content/uploads/HART-Attack.pdf

Milivojevic, S. (2021) *Crime and Punishment in the Future of Internet: Digital frontier technologies and criminology in the twenty-first century*, Abingdon and New York: Routledge.

Molnar, P. (2019) 'Technology on the margins: AI and global migration management from a human rights perspective', *Cambridge International Law Journal*, 8 (2): 305–330.

Molnar, P. (2021a) 'Technological testing grounds and surveillance sandboxes: migration and border technology at the frontiers', *Fletcher Forum of World Affairs*, 45 (2): 109.

Molnar, P. (2021b) 'Inside new refugee camp like a "prison": Greece and other countries prioritize surveillance over human rights', *Conversation UK*, 27 September.

Molnar, P. and Gill, L. (2018) 'Bots at the gate: a human rights analysis of automated decision-making in Canada's immigration and refugee system', Citizen Lab and International Human Rights Program, citizenlab.ca/wp-content/uploads/2018/09/IHRP-Automated-Systems-Report-Web-V2.pdf

Monroy, M. (2022) 'Meeting in Washington: EU plans biometric super database', 20 October, digit.site36.net/2022/10/20/meeting-in-washington-eu-plans-biometric-super-database

Morozov, E. (2022) 'Critique of techno-feudal reason', *New Left Review*, 133/134.

Noble, S. U. (2018) *Algorithms of Oppression: How search engines reinforce racism*, New York: NYU Press.

Nonnecke, B. and Dawson, P. (2022) 'Human rights impact assessments for AI: analysis and recommendations', *Access Now*, October, https://www.accessnow.org/wp-content/uploads/2022/11/Access-Now-Version-Human-Rights-Implications-of-Algorithmic-Impact-Assessments_-Priority-Recommendations-to-Guide-Effective-Development-and-Use.pdf

O'Neil, C. (2017) *Weapons of Math Destruction: How big data increases inequality and threatens democracy*, New York: Penguin.

Palvia, P., Baqir, N. and Nemati, H. (2018) 'ICT for socio-economic development: a citizens' perspective', *Information and Management*, 55 (2): 160–176.

Parker, B. (2019) 'New UN deal with data mining firm Palantir raises protection concerns', *The New Humanitarian*, thenewhumanitarian.org/news/2019/02/05/un-palantir-deal-data-mining-protection-concerns-wfp

People & Planet (2021) *Divest Borders Action Guide*, https://peopleandplanet.org/system/files/resources/Divest%20Borders%20Action%20Guide.pdf

Prentzas, G. (2021) 'The EU, the externalisation of migration control, and ID systems: here's what's happening and what needs to change', Privacy International, privacyinternational.org/long-read/4651/eu-externalisation-migration-control-and-id-systems-heres-whats-happening-and-what

Privacy International (2016) The Global Surveillance Industry, July.

Privacy International (2019) *Surveillance Company Cellebrite Finds a New Exploit: Spying on asylum seekers*, 3 April.

Privacy International (and No Tech for Tyrants) (2020) *All Roads Lead to Palantir: A review of how the data analytics company has embedded itself throughout the UK*, September.

Privacy International (2021) *The UK's Privatised Migration Surveillance Regime: A rough guide for civil society*, February.

Privacy International (2022) *Buddi Limited – Immigration enforcement's favourite tracking buddy*, November.

Pulignano, V., Meardi, G. and Doerflinger, N. (2015) 'Trade unions and labour market dualisation: a comparison of policies and attitudes towards agency and migrant workers in Germany and Belgium', *Work, Employment & Society*, 29 (5): 808–825.

Raghuram, P. (2009) 'Which migration, what development? Unsettling the edifice of migration and development', *Population, Space and Place*, 15 (2): 103–117.

ResponsibleData (2019) 'Open letter to WFP re: Palantir Agreement', responsibledata.io/2019/02/08/open-letter-to-wfp-re-palantir-agreement

Riotta, C. (2020) 'Trump administration bought access to cell phone database to target immigrants, report says', *The Independent*, independent.co.uk/news/world/282mericas/trump-ice-immigration-cellphone-location-data-report-surveillance-tracking-a9324221.html

Rohrlich, J. (2020) 'Court document shows US troops surveilling migrants at the Mexico border', *Quartz*, qz.com/1815249/us-troops-are-surveilling-migrants-along-the-border-with-mexico

Sadowski, J. (2019) 'When data is capital: datafication, accumulation, and extraction', *Big Data & Society*, January–June: 1–12.

Sánchez-Monedero, J. and Dencik, L. (2020) 'The politics of deceptive borders: "biomarkers of deceit" and the case of iBorderCtrl', *Information, Communication & Society*, 25 (3).

Schulz, C. (2022) *Anticipating Cyberbodies Offshore: Europe's racio-technologized frontiers*, unpublished thesis, MSc Migration Studies, University of Oxford.

Slavin, A. (2021) 'Digital identification for the vulnerable: continuities across a century of identification technologies', in E. E. Korkmaz (ed) *Digital Identity, Virtual Borders and Social Media*, Cheltenham: Edward Elgar Publishing.

Solano, G. (2022) 'Indicators and survey data to understand migration and integration policy frameworks and trends in the EU', in A. A. Salah, E. E. Korkmaz and T. Bircan (eds) *Data Science for Migration and Mobility*, Oxford: Oxford University Press.

Soprano (2022) *Digital Interactions and Transformation for Financial Institutions*, Santa Clara: Frost & Sullivan.

Talbot, R. (2021) 'Automating occupation: international humanitarian and human rights law implications of the deployment of facial recognition technologies in the occupied Palestinian territory', *International Review of the Red Cross*, international-review.icrc.org/articles/ihl-hr-facial-recognition-technology-occupied-palestinian-territory-914

Taylor, L. and Meissner, F. (2020) 'A crisis of opportunity: market-making, big data, and the consolidation of migration as risk', *Antipode*, 52 (1): 270–290.

Tazzioli, M. (2021) 'The technological obstructions of asylum: asylum seekers as forced techno-users and governing through disorientation', *Security Dialogue*, 1–18.

Thapa, S. (2022) *Affective Nepali Student Journeys: Unearthing the racial contours of the UK student visa application process*, unpublished thesis, MSc Migration Studies, University of Oxford.

Toyama, K. (2011) 'Technology as amplifier in international development', in Proceedings of the 2011 iConference, ACM, pp 75–82.

United Nations (2018) *Global Compact for Safe, Orderly and Regular Migration: Final draft*, un.org/pga/72/wp-content/uploads/sites/51/2018/07/migration.pdf

Van der Ploeg, I. (1999) 'The illegal body: "Eurodac" and the politics of biometric identification', *Ethics and Information Technology*, 1: 295–302.

Véliz, C. (2020) *Privacy is Power: Why and how you should take back control of your data*, London: Bantam Books.

Vukov, T. and Sheller, M. (2013) 'Border work: surveillant assemblages, virtual fences, and tactical counter-media', *Social Semiotics*, 23 (2): 225–241.

Weber, I. (2022) 'Using Facebook and LinkedIn data to study mobility', in A. A. Salah, E. E. Korkmaz and T. Bircan (eds) *Data Science for Migration and Mobility*, Oxford: Oxford University Press.

Weitzberg, K., Cheesman, M., Martin, A. and Schoemaker, E. (2021) 'Between surveillance and recognition: rethinking digital identity in aid', *Big Data & Society*, 8 (1): 1–7.

World Food Programme (2019a) 'Palantir and WFP partner to help transform global humanitarian delivery', wfp.org/news/palantir-and-wfp-partner-help-transform-global-humanitarian-delivery, 5 February.

World Food Programme (2019b) 'World Food Programme begins partial suspension of aid in Yemen', 20 June, wfp.org/news/world-food-programme-begins-partial-suspension-aid-yemen

Zelinsky, W. (1971) 'The hypothesis of the mobility transition', *Geographical Review*, 61 (2): 219–249.

Zuboff, S. (2019) *The Age of Surveillance Capitalism: The fight for a human future at the new frontier of power*, London: Profile Books.

Index

A

academics 2, 11, 26, 34, 35, 59, 69
   debates on data and migration 57–58
   migrant academics in UK 120
Accentura 74
accountability 3, 36, 44, 111, 127, 129
advertising 32, 33, 37–39, 46, 50, 71
Aegean Sea, migrants in the 121, 125
Afghanistan 53, 56, 61, 66, 74, 92
   facial recognition software 105
   refugees and work 26, 27, 28
Africa 59, 67, 69, 79, 80, 81, 84
agenda-setting 21, 87, 88, 101, 113
AI algorithms 4, 7, 8, 43, 63, 84, 106, 120, 121
   and data analysis 23, 46, 104
   lie detectors 9, 31, 75, 80, 101, 117
   for processing applications 30, 69–70
Airbus 17, 73, 74, 105
Alan Turing Institute 112
Ali Salah, Albert 58
alternative approaches 3, 44, 57, 107
   alternative technology 129–131, 132
   in digital identity 88, 101, 112
Amazon 3, 34, 55, 71, 73, 75, 101, 113
Anduril 73, 75
Ankara Agreement 26
Armenia-Azerbaijan war 77
Australia 11, 17, 19, 66, 92
authoritarianism/dictatorships 2, 18, 71, 117, 130
automatic number plate recognition (ANPR) 69
Avatar 75

B

BAE systems 73
Bangladesh 11, 17, 47, 53, 55
Bauman, Zygmunt 40, 43
behavioural surplus 39–40
Biden, President 19, 73
big data analysis *see* data analysis
Big Tech *see* technology companies
biometric data 13, 29, 62, 68, 100, 103–106, 120

Bircan, Tuba 58
Bitcoin 100, 108–109
Black people 8, 9, 105, 107
   and facial recognition 80
   and oppression in the US 4–5
Black Sea states, refugees from 27
blockchain technology 64, 105–106, 107–111
boats 73, 80, 82, 121, 123, 125, 130
borders and surveillance technologies, 19, 28, 42, 44, 48, 68, 88
   about border management 1–4, 117, 118, 119, 120
   analysing before arrival at 17
   border agents accessing mobile phones 40
   border externalization 81, 84, 118
   border security 2, 129
   border wall policies 10, 19, 78
   cross-border monitoring 120
   migrant-oriented approach to border control 79
   militarization of 11, 68, 72, 125
   and right to asylum 82–83
   stakeholders in management of 20–21, 23
   and surveillance technologies 10–11, 17–19
   see also humanitarian aid; smart borders; surveillance capitalism
Brexit 33, 66, 121, 131
buffer zones 81

C

Cambridge Analytics 66
Canada 11, 17, 19, 75, 92, 122
   and visa scanning 69
capitalism 3, 17–19, 43–45
   contemporary capitalism 44, 70, 116
   and political policies 121–122
   rentier capitalism 26
   and selection for labour 12
   trajectory of 17, 18, 31, 44, 113, 119
   and the working class 24–29
   see also migration and capitalism; surveillance capitalism
Caribbean, labour from the 21
cash aid 94

INDEX

Central American refugees 61
Central Asia, refugees from 27, 59
Cheesman, M. 112
children 75, 90, 91
China 11, 54
CIA 54, 74, 76
class 18, 19, 32, 60, 70, 124, 131
　discrimination in education 8
　selection on basis of class 30
Clearview AI 104
climate crisis 18, 19
cloud storage 46, 73, 76, 104
colonialism 72, 79–80, 85, 127
　attitudes and technologies 78, 101, 111
　colonial relations 16, 18, 21, 50, 89
companies *see* financial firms; military companies; also private companies; technology companies
consent 41, 49, 53, 62–64, 67, 80, 113–114
　and facial recognition 104
　relevance of 126–128
construction industry 23, 24, 27, 29, 63, 92, 119, 122
consumers, becoming 29, 45, 92–93, 126
countries, dependent and weak 49–50, 72, 102
COVID-19 4
crime, predicting 4, 9, 89

D

data
　academic debates about data 57–58
　access to data of millions of people 23
　anonymization 57, 62–63
　cloud storage 46
　data minimization 111–115
　data sharing 47, 57, 91
　hacking of 56, 78, 105, 129
　historical data sets 48
　and humanitarian agencies 52–55, 55–57
　from other countries 29–31
　quality and diversity of datasets 5
　real-time data 49
　security 56, 115
data analysis 8, 46, 119
　and dominance by companies 46–48
　and ethical issues 62–63
　and future forecasting 32
　and manipulation 33
　and migration movements 58–60
　and migration scholars 59–60
　and surveillance capitalism 67
　see also AI algorithms
data collection 12–13, 99, 102, 112, 126
　comprehensive 47–48
　and humanitarian agencies 52–55, 55–57
　and privacy 100
deaths and threats to life 19, 47, 54, 70, 87, 100, 129
　and dangerous crossings 73

decentralized systems 105, 109, 110, 112
democracy 44, 66, 72, 117
Democratic Party 10, 19
deportation 28, 83, 124
deregulation 43, 127
development projects 83–85, 93
dictatorships *see* authoritarianism/dictatorships
digital identity 23, 34, 43, 119, 125
　controversy over 112–113
　hierarchies of 98–99
　importance for aid 88–92, 95–96
　and integration of refugees 92–98
　and recording entire lives 97
　self-sovereign identity 112
　supportive organizations of 102–105
　and surveillance capitalism 87–88, 96–98, 98–111
digital inequality 60–63
digital literacy *see* literacy
discrimination 4–5, 88, 112
DNA tests and analysis 62, 76, 104
documentation 26, 29, 65, 89, 90
donor countries 18, 21, 36, 51, 52, 116
　and blockchain technology 106
　and development projects 83–85
　and digital identities for aid 95
drones 10, 48, 49, 73, 77
drugs, illegal 35
Dublin Regulation 83

E

education and training 25, 84, 89, 90, 92, 97
　school exams 4–5, 8–9
Elbit 35, 68
emergency situations 2, 53, 93–94, 109
　and identity registration 90
employment *see* work
English channel boats 73
Erdoğan, President 19, 27, 122
errors, system 78, 128–129
Ethereum 109
ethical concerns 58
　and data analysis 60–63
Ethiopia 16, 67
ethnicity 74, 80, 88, 102, 131
　ethnic discrimination 112
European Space Agency (ESA) 65
European Union 11, 17, 19, 59, 124
　and biometric data 104, 105
　blockchain projects by 110
　and digital identity 102
　EUAA predicting migration 67
　Eurodac report errors 129
Frontex 35, 73
funding of warlords 80
funding refugee camps 72
　and human rights 12
　preventing migration from Africa 81

and privacy 41
Readmission Agreement (Turkey) 27, 81
resettlement programmes 83
and Ukrainian refugees 25
exam algorithm 4–5, 8–9
exploitation 16, 22, 130
eye scanning 64, 96, 104

F

Facebook *see* Meta
facial recognition 40, 68, 71, 75, 80, 104, 117
  and low-accuracy rates 74
fake news 1, 66
families 90, 111
  decision to migrate 59
  reunion visas 13, 30
financial firms 17, 47, 48, 88, 117
  and digital identity 102
  and lobbying 96
  and mobile banking 39, 95
  projects with humanitarian aid 36, 97, 107, 120
  and real-time data 46
fingerprints 64, 104, 105, 106
food insecurities 54, 77
forced migration 18, 19, 89, 90, 93
foundational identities 99
free movement of labour 19–22
French government 105
Frontex 35, 79, 104
  and drones 73
funding and investment 17, 31, 51, 76, 112, 118–119
  of agencies and academia 21
  for innovative projects 106–107
  of warlords 80
Future Borders and Immigration Systems 73

G

Gaza and product testing 72
gender 16, 60, 85
Germany 29, 70, 92, 106
  and guest-worker programme 21
  and selective migration 26
  and Syrian refugees 25
gig economy 5–6, 84–85, 123
Global Compact for Migration 20
Global North 11, 42, 49, 114
Global South 11, 71, 85, 100, 111, 114
  as host countries 49
  workers from the 16
globalization 21
goat pricing 67
Google 3, 32, 34, 47, 71, 75
Greece 19, 21, 60, 69, 83, 125
  Greek island camps 11, 31, 40, 72, 80
  and push-back policy 79
Gulf countries 29, 92

H

H&M 17, 28
hacking 56, 78, 105, 129
Haiti 53, 61
healthcare 58, 90, 91, 92
Hewlett-Packard 74
Homeland Security 73, 104
Houthi government 54, 55
HSBC 47
humanitarian aid agencies 4, 18, 34, 41, 116, 118
  cash aid 94
  and consent and privacy 127
  and data organizations 52–55, 55–57, 67
  and digital identity 88–92, 95–96
  humanitarian-development nexus 24, 93
  innovative/short-term projects 45, 56–57, 106–107
  reliance on tech companies 51
  see also United Nations

I

IBM 34, 55, 74
iBorderCtrl poject 75
identity, official 88–91, 102
imperialism 76, 78
India 28, 92
inequality 3, 5, 7, 18, 61, 111
  digital inequalities 60–63, 130
information, access to 130–131
infrastructure 38, 76, 89, 93, 107
innovative projects 45, 106–107
In-Q-Tel 74
institutions 1, 44, 66
International Organization for Migration (IOM) 21
internet access 84, 89, 92, 95, 107, 112, 129
Iranian refugees 27, 77
Iraq 27, 74, 90
  facial recognition software 105
irregular migrants 27, 82
ISIS 90
Israel 72, 73
  Elbit security company 35, 68
Istanbul Metropolitan Municipality (IBB) project 110
Italy 69, 79

J

Jamaica 122
Japan 11, 106
Johnson, Boris 19, 44, 121
Jordan 24, 25, 53, 84
  Zataari camp 64, 96, 105, 109

K

Kenya 53, 107
Kurds in Syria 89

INDEX

L

labour 19, 26, 60, 119, 122–123
  cheap labour 28, 92, 123
  free movement of 19–22
  guest-workers and former colonies 21
  joining the labour force 92, 123–124
  market defined policies 120
  market selection of migrants 12–13
  and migration policies 22–24
  unions 5, 16, 22, 28, 39
  see also migration and capitalism; work
language training 24, 25, 30
Latin America 59, 66, 91
leather work 27
Lebanon 53
legal frameworks 2, 22, 44, 56, 121
  for asylum and refugees 20, 82
  for consent and privacy 126–128
  for digital technologies 110–111
  law enforcement 4–5, 49
legal migrants 13, 22, 49, 69, 117, 122
  documentation requirements 29
Leggeri, Fabrice 79
Libya 69, 70
  slave camps in 80
lie detectors 9, 10, 30, 40, 75, 101, 120
literacy 64, 109
  digital literacy 84, 108, 111, 112
lobbying 34, 35, 73, 74, 76, 96, 102, 113
Lockheed Martin 74

M

Mali 105
Malta 69, 70, 79
Manchester Metropolitan University 75
marketing 37–38
Marxist theories 16, 19
Mastercard 47, 102, 107
May, Theresa 120
Mediterranean migrants 18, 69, 70
Meta 3, 47, 71
metal work 25
metaphysical consequences of technology 7–10
Mexico 11, 17, 75
  US-Mexico border 10, 19, 69, 73, 101, 104
Microsoft 34, 55, 74, 100, 102, 107, 113
  and cloud services 75
Middle East 67, 69, 80
migrants and asylum seekers 19, 121
  attitudes towards 24, 45, 51, 59, 62
  categorization of 13, 19–20, 24, 32
  deaths/threats to life 19, 47, 54, 70, 73, 87, 100, 129
  documentation 26, 29, 65, 89, 90
  errors and delays in processing 129
  irregular migrants 27, 82
  legal migrants 13, 22, 29, 49, 69, 117, 122

refugee camps 11, 24, 31, 40, 57, 69, 70, 72, 80
  refugees in Turkey 27–29
  rights of 11, 12, 73, 82–83
  and their problems 25, 59, 85
  and threat of deportation 28
  unable to influence technologies 114
  see also consent; digital identity; mobile phones; privacy; surveillance technologies; work
migration and border management
  concept of resilience 83–86
  and data 48–52
  detecting/predicting movement 11, 32, 58–60, 65, 66–67, 68, 73, 117
  integration of migrants 92–98
  and mass migration 22, 49, 70, 71, 90–91
  migration policies 22–24, 120
  push-back policies 73, 79
  refugee burden sharing 83
  resettlement programmes 25, 83
  security-oriented approach to 35, 48
  selecting workers 24–28
  selection and monitoring of migrants 12–15
  see also borders; data analysis; digital identity; surveillance technologies
migration and capitalism 11, 16–19, 17, 44–45
  data from other countries 29–31
  free movement of labour 19–22
  and the working class 24–29
migration studies 18, 59, 87
military companies 11, 14, 17, 33, 36, 40, 73–74
  and tech companies crossover 76–77
  testing of technologies 119
minority groups 80, 89
misinformation 66
mixed-motive migration 18, 20
mobile banking 23, 50, 52, 91, 96, 96–97
  bank accounts 24, 39, 89, 92, 95, 125
mobile phones 23, 40, 68, 94
  access to 60–61
  alternative approaches to apps 130
  confiscation of 66, 80
  phone operators 14, 20, 34, 46, 103
  smartphones 49, 50, 53, 60, 80, 81, 95, 125
money laundering 48, 89
monitoring migrants 10, 17, 27, 65, 72, 85, 92, 105
monopolies, global 33, 34, 43, 51
Morocco 81
movement detection 68, 73, 117
Myanmar 89

N

natural disasters 89, 90, 93
  and business opportunities 53

149

neoliberalism 21, 43, 121, 127
Nepal 28, 61, 92, 122
  blockchain projects in 107, 109
  view of Nepali people 26
new customers/markets 53, 96, 125
newspapers online 98, 99
Next 28
NGOs 4, 18, 29, 35, 56, 100, 127, 130
  and alternative tech solutions 132
  resettlement programmes 25
Nigeria 53
  blockchain projects in 107
no harm, principle of 47, 63, 67, 92, 100, 131
Northrop Grumman 73, 104
Norway 69, 70, 106

O

Obama, President 73
oligarchic structure 11, 12, 46, 76, 88, 116
  and cooperation between companies 51
  and surveillance capitalism 33–37
oppression 2, 40, 130
Orange 47, 102

P

Pacific Islands, migrants from 92
Pakistani refugees 27, 28
Palantir 34, 54, 55, 74–75, 107, 118
Palestinian people 35
  and testing of products 72
Pegasus attack 77
Pentagon 54, 55, 76
permanent residence permit 26
personal data 13, 35, 58
personnel changes, organizational 34, 74, 76, 116
Philippines 26, 59, 92, 122, 129
  migrant females 16
Poland 60
police 4, 71, 74, 101, 104
politics 18, 19, 21, 39, 54, 71, 89, 130
  political solutions 3, 131
Portugal 73
poverty 8, 18, 26, 49–50, 51, 60, 61
  in Turkey 28
predicting migration 66–67, 117
Primark 28
prison populations 5, 9
privacy 41, 49–50, 63–66, 113, 114–115
  relevance of 126–128
  and smart borders 80–82
  and surveillance capitalism 40–43
Privacy International 73
Privacy is Power 41, 65
private companies 3–4, 33–34, 118
  agenda-setting 21, 23, 88
  in competition in countries 125
  and digital identity 97
  and dominance over data 46–48
  and funding for projects 106–107
  and personnel changes 34, 74, 76, 116
  and political policies 122
  and privacy and consent 128
  role of in border control 20–21, 72
  see also oligarchic structure; private companies
private schools 5, 8
product development and design 23, 45, 126–127
profits 23, 77, 96
Project Maven, virtual-reality system 75
protest, right to 71, 104
public authorities 52, 110–111
push-back policies 73, 79

R

racism 16, 18, 25, 122
  racial capitalism 78, 85
RAVEn programme 73
Readmission Agreement 27
real-time data 46, 49, 68, 71, 74
Red Cross 56, 91
reform-oriented approach 4–6, 112, 126–127
Refugee Convention 1951 20
registration and verification processes 53
regular/irregular migration 19
remittances 47, 50, 59
Renault 17
rentier capitalism 37, 38
Republican Party 10
resettlement programmes 25, 83
residence permits 13, 49
resilience, concept of 83–86
rights 11, 12, 49, 66, 73, 89, 121, 127
  privacy rights 65, 80–82
  right to asylum 82–83
  right to protest 71, 104
right-wing politics 18, 54
risk scores 9, 49
robots 6, 10, 48, 77
Rohingya people 53, 55, 89
Romania 26
Russia 11, 59, 77, 91, 119

S

'safe country,' concept of 83
Safran 74, 105
satellites and sensors 11, 48, 49, 65–66, 68
Saudi Arabia 17, 55
Schengen area 13, 29, 104
Schulz, C. 79
Sea-Watch 130
secrecy 36, 70, 117
security-oriented approach 33, 36, 47, 48, 117, 118

INDEX

selecting workers 22, 24–29
  on basis of class 30
Senegal 105
short-term projects 107
Silicon Valley companies *see* technology companies
Skype 129
slavery 16, 70, 78, 80
Slavin, A. 112
small number of companies *see* oligarchic structure
smart borders 11, 82–83
  and academic projects 75
  developing and testing technologies 72–74
  in different regions 69–72
  and digital identity 43, 92
  and privacy 78–79, 80–82
  technologies 68–69
  and threats to peace 77
smugglers 81, 82, 130
social control 71
social media 37, 39, 46–47, 50, 51, 60, 68, 71
  AI and social media mining 104
  alternative groups 130
  and manipulation 66
society, threats to 4, 10, 31, 42, 68, 77, 119, 132
software engineers 2, 24, 26, 92, 122
Somalia 67, 91
South American refugees 61
South Asian refugees 21, 80
sovereign rights of states 41
  and privacy 81
Spain 21
structural analysis 4, 6, 18
students 13, 19, 30, 49, 69
supply chains of capitalist companies 28, 85, 92, 123
surplus value 22, 32, 38, 93
surveillance capitalism 6, 12, 43–45, 76, 121
  and agenda-setting 88
  and data analysis 67
  development of 31–33
  and digital identity 87–88, 96–98, 98–111, 115
  as an oligarchic structure 33–37
  and privacy 40–43, 65
  rejection of 131–132
  and Zuboff's critique 37–40
surveillance technologies 10–11, 17, 48–49, 72–74, 123
  biometric data 13, 29, 62, 68, 103–106, 120
  consequences of 10, 132
  DNA tests and analysis 62, 76, 104
  drones 10, 48, 49, 73, 77
  eye scanning/fingerprinting 64, 104, 105, 106
  facial recognition 40, 68, 71, 74, 75, 80, 104, 117

lie detectors 9, 10, 30, 40, 75, 101, 120
  opposing surveillance technologies 78–80
  robots 6, 10, 48, 77
  satellites and sensors 11, 48, 49, 65–66, 68
  see also AI algorithms; data analysis; digital identity; monitoring migrants; smart borders
Sweden 69, 83, 106
Syrian refugees 27, 28, 53, 92

T

Taliban 56, 91
technological change/development 7, 48
  alternative approaches 132
  beneficiaries of 17–18
  consequences of technologies 6, 7–10, 119, 132
  critical approaches to 4–5
  developing and testing 72–74
  technological solutionism 2, 68
technology companies 2, 11, 23, 29, 37
  and academic projects 35
  attitude to regulation 42
  business models 32–33, 46, 50, 56–57, 97, 113
  cooperation with military/security industry 36–38, 69, 70–71, 76–77, 87, 124
  and data analysis 32–33, 67
  as global monopolies 6, 34, 43, 104
  and humanitarian agencies 52–55, 95–96, 103, 107
  investments of 13, 116, 121
  and secrecy 117
  and surveillance capitalism 76
  testing technologies 23, 35, 40, 42, 100, 117, 124
  see also data analysis; digital identity; surveillance technologies
Tekever 73
telecoms companies 34, 46, 50, 76, 95, 96, 107
  and digital identity 102
temporary protection status 20
terrorism 35, 48, 77, 89
textile/garment work 17, 24, 27, 28, 85, 92, 123
Thailand, blockchain projects in 107
Thales 17, 74
Thapa, Shuvashish 26
Thiel, Peter 74
tourism 13, 30, 49, 69, 92
Trump, President 10, 19, 33, 54, 66, 73, 74, 122
Tunisia 81
Turkey 11, 17, 21, 26–29, 36, 53, 81, 83, 92
  and Aegean Sea migrants 125
  blockchain projects in 107, 110
  e-government services 99
  EU funds to retain migrants 27

and identity registration 90
migrants joining the labour force 123–124
and small-scale workshops 24
Turkish migrants to UK 26–27
Turkmenistan 28

U

Uber 85
Uganda, suppression of opposition in 71
Ukraine 26, 67, 77, 125
  official documentation 91
  refugees from 25, 60
United Kingdom 11, 35, 83, 92, 106, 112, 120, 122
  and Ankara Agreement 26
  Brexit 66, 121
  contracts for surveillance 73, 74–76
  and digital identity 102
  exam algorithm 4–5, 8–9
  labour from former colonies 21
  and migrant flows 15, 121, 124
  political campaigns in 33
  and selective migration 17, 26, 92, 120
  surveillance of legal migrants 131
United Nations 4, 18, 44, 54, 102, 118–119
  definitions of migrants 20
  development projects 83–84
  and identity registration 90
  and reliance on tech companies 52, 55
  resettlement programmes 83
  Sustainable Development Goals (SDGs) 89, 102
  UNHCR 21, 67, 91
United States 11, 17, 66, 92, 106, 124
  and biometric data 104
  and Black people 4–5
  border wall policies 10, 19, 72–73
  and facial recognition 74
  political campaigns in 33
  resettlement programmes 25
  and surveillance contracts 73–75
  US-Mexico border 69, 73, 101, 104, 125
University of Oxford 101

V

Véliz, Carissa 36, 41, 42, 65
Venezuela 91

virtual wall policy 19, 78
virtual-reality systems 75
visas 13, 26, 30, 49, 65, 69, 120
  AI algorithm analysis 70
  overstaying 82
Vodafone 47, 102, 107
vulnerable people 42, 47, 55, 61, 91, 93, 100

W

wages, reducing 22, 25, 27, 28, 45, 131
wars and conflicts 2, 18, 22, 53, 55, 90, 93
  future risk of war 77–78, 117, 124
white people 5, 78
women 16, 91, 94, 105, 111, 129
  and access to digital tools 60
work 23, 24, 27, 28, 29, 59, 92, 119, 122–123
  agricultural work 24, 26, 27, 123
  catering work 85, 123
  childcare 16, 122
  for displaced people 84–85
  domestic service 16, 26, 28, 59, 92, 123, 129
  healthcare work 84, 92, 122
  precarious work 6, 16, 120
  professional work 123, 131
  work permits 13, 30, 92
  working conditions 5, 27
working class 6, 12, 122
  international working class 22–24
  selection of the 24–29
  shaped and positioned 28, 45, 48, 70, 82, 92, 120
World Bank 102, 115
World Economic Forum 102, 115
World Food Programme (WFP) 21, 54, 55, 74, 91

Y

Yemen 26, 28, 47, 53, 54, 55, 125
YouTube 66

Z

Zara 17, 28, 124
Zuboff, S. 37–40

www.ingramcontent.com/pod-product-compliance
Lightning Source LLC
Chambersburg PA
CBHW071712020426
42333CB00017B/2238